Designing the Coal Preparation Plant of the Future

Designing the Coal Preparation Plant of the Future

Editor

Minh Chi Ngo

scitus academics

Designing the Coal Preparation Plant of the Future
Edited by **Minh Chi Ngo**

ISBN: 978-1-68117-572-0

Library of Congress Control Number: 2016943007

SCITUS Academics LLC,
www.scitusacademics.com
Box No. 4766, 616 Corporate Way,
Suite 2, Valley Cottage,
NY 10989

Printed in United States of America on Acid Free Paper

Preface

Coal preparation plants are designed to clean impurities, such as soil, sand and rock from coal, to ready it for transportation to power plants and coking facilities for use in steel processing. Monitoring flow through the coal preparation process can be difficult as the coal is abrasive and can cause inaccuracies and maintenance issues for intrusive instruments. A Coal preparation plant generally use gravity process equipment to separate the refuse from the product (coal). Coal has a specific gravity between 1.35 and 1.5, while the refuse rock has a Specific Gravity of 2.1 to 2.3. Heavy Media is the most popular method of cleaning coarse sizes. Heavy media cyclones are being used more often for fines size fractions. Flotation is generally used to clean the -28 mesh size fraction, although spirals and heavy media cyclones have shown success in cleaning down to 100 mesh coal feed. Spirals are generally used for middling sizes.Modular Coal Preparation Plants (MCPP) prepares coal for a cleaner burning process, enhanced calorific value and cleaner emissions. MCPPs consist of several modules including a raw coal preparation (crushing and screening station), a coal feed and washing module (jigging and Heavy Medium cyclone) as well as a dewatering module. *Designing the Coal Preparation Plant of the Future*provides comprehensive overview on several current and future challenges that must be considered in the design and operation of coal preparation facilities. These challenges dictate that we constantly strive to improve plant. Coal Preparation Engineers are meeting this challenge of the future by designing large modem environmentally acceptable plants.

Table of Contents

1

CHAPTER

Design of Integrated Manufacturing System of Coal Preparation Plant Basedon Multi-Agent

Deng Jian-jun[a,*], Huang Jian-yu[b], Kuang Ya-li[a]

[a] *School of Chemical Engineering and Technology, China University of Mining & Technology, Xuzhou, Jiangsu 221116, China*

[b] *Huaibei Mining(Group)Co., Ltd, Huaibei, Anhui, 235000, China*

ABSTRACT

Production process of coal preparation plant is a complicated process of multi-layer, which is determined by the production characteristics. The emphasis of information construction in coal preparation plant is the automatic data collection and data analysis. This paper constructs the integrated manufacturing system of coal preparation plant based on multi-agent, including the system analysis, module division, module construction, Agent construction, and software implementation. The production automation system has been successfully integrated with the information management, providing a useful auxiliary decision support to the production management in practice.

Keywords: Multi-agent; coal preparation plant; integrated manufacturing system

1. INTRODUCTION

Coal preparation plant is an enterprise where coal products of various quality and specifications are graded after removing impurities from raw coals. The whole production process of coal preparation plant is a complicated system of multi-layer and multi-structure, which integrates production control, operation and management and technology, with coal as the processing target. It is characterized by the close link of different operating processes, high level of mechanization and automation and continuity and short process[1-2]. The integrated manufacturing system of coal preparation plant aims at the successful integration of dispersed control system, scheduling system and management decision-making system, with the acquisition of all information required in the production process. With the application of automation technology, information technology, computer technology, minerals processing technology, and modern management science, an integrated manufacturing system characterized by excellent adaptation to various production environments and markets demand, optimal, superior quality, high profit, and high flexibility is expected to construct. According to the author's experience in studying the information construction of coal preparation plant over years, five problems as follows exist in the current integrated manufacturing system:

1) The production and operation of coal preparation plant should take initiative to meet the change of external environments. Otherwise, the survival and development of them will be a great challenge[3].

2) The automation of a single machine and the centralization of technical process control require well-organized and soundly-planned automation technical transform.

3) As a flow type enterprise, the automatic collection of data and analysis and extraction of effective information should be the emphasis of the construction.

4) Intelligent motoring of production equipment operation, reasonable optimization of process parameters and the

measurement and control of product quality should be realized in the production process.

5) A good compatibility to the existing system should be guaranteed. In recent years, remarkable resultss have been achieved in automation and information management of coal preparation plant at home and abroad. As a result, the new system must be an open one.

2. INTEGRATED MANUFACTURING SYSTEM MODULE OF COAL PREPARATION PLANT BASED ON MULTI-AGENT

Nowadays, the production and operation of coal preparation plant are increasingly market-oriented and networked. The major functional departments and cooperating enterprises are geographically distributed and autonomous, which demands the resource coordination and cooperation to realize the aim of the system. According to the internal distribution of software and hardware resources, in the present study, the integrated manufacturing system of coal preparation plants is regarded as a system consisting of several interactive intelligent Agents, with one application corresponding to an Agent.

The integrated manufacturing system module designed in the present study is shown in Fig. 1[4]. The module consisted of several static and dynamic Agents. Communication between different Agents is conducted by KQML/XML. In order to solve the problems such as compatibility of application system and supporting platform, friendly interaction between human and machine etc., five Agents are designed in the system. The five Agents can communicate with each other through the proxy Agents and ultimately interact with the users[5-6].

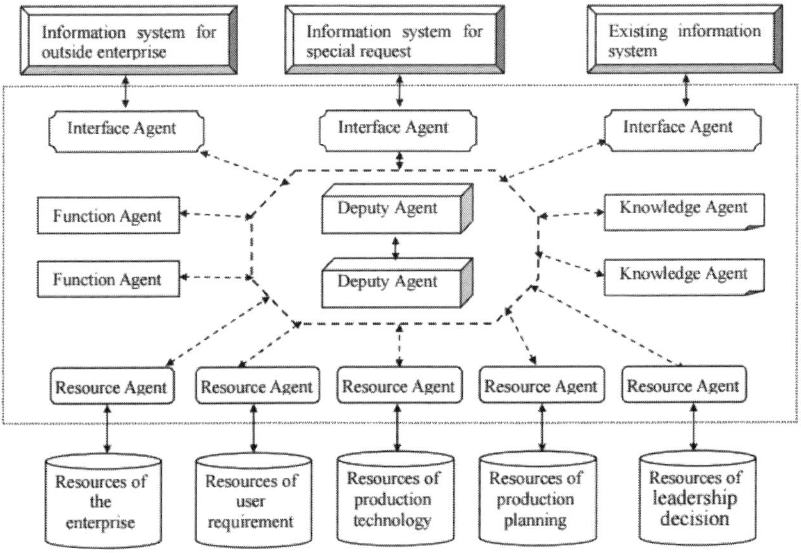

Figure 1. The integrated manufacturing system module of coal preparation plant based on multi-agent

3. REALIZATION OF INTEGRATED MANUFACTURING SYSTEM

Based on the module shown in Fig.1, different types of function Agents were designed. In essence, the integrated manufacturing system is composed of many links bearing different tasks. The links cooperate to fulfill the task of coal processing while fulfilling their respective tasks.

Different types of Agents are decomposed in the present study. There are two modules of decomposition: function decomposition and entity decomposition [7]. The former involves the packaging of some functional modules with Agent, which has no direct corresponding relationship with the entity. While the latter involves the actual existing entity such as raw coal, the workers, the operation and so on. Thus Agent has a definite relationship with the entity. In the present module (as shown in Fig. 2, entity decomposition is mainly applied on the basis of function decomposition.

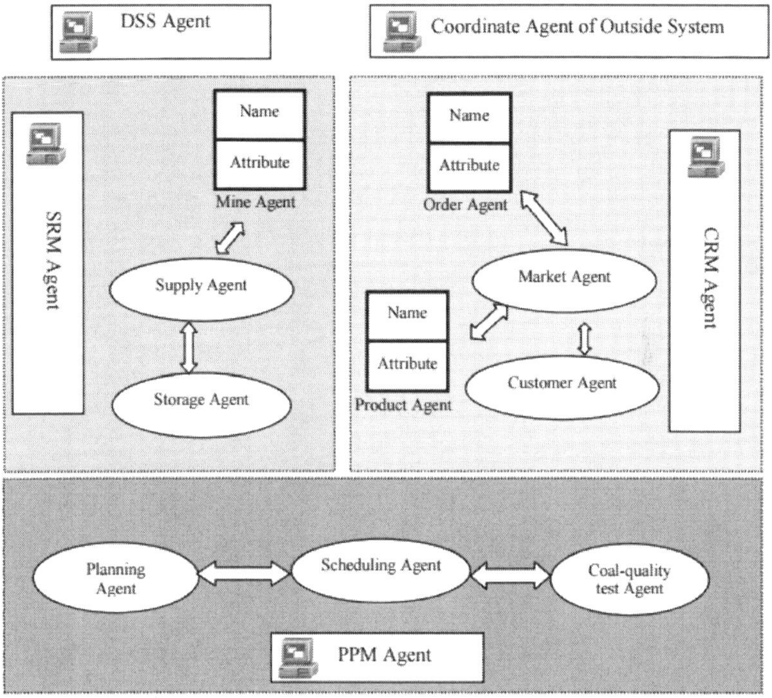

Figure 2. Agent decomposition schematic diagram

Concerning the production and operation features of the coal mine preparation plant, four major management systems were designed according to different functions. They are: supplier relationship management system (SRM), customer relationship management system (CRM), production process management system (PPM), and decision support system (DSS). Corresponding to each system, intelligent Agents ranging of SRM, CRM, PPM, and DSS were designed respectively. In each intelligent Agent, relevant function and entity decompositions were set. In order to create a harmonious internal and external environment for the integrated manufacturing of coal preparation plant, a coordination intelligent Agent of external system was also set.

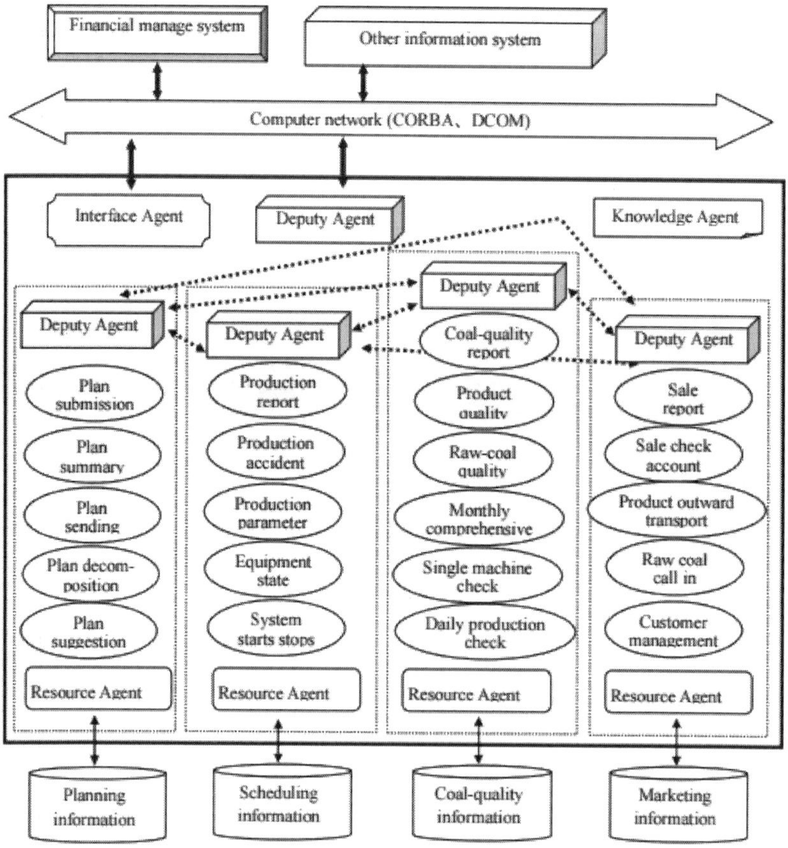

Figure 3. Architecture of the intelligent agent for coal preparation production

4. CASE STUDY

The application of the designed system can be seen in the case of the production and management system of LinHuan Coal Preparation Plant of Huaibei Mining(Group)Co., Ltd, where the production and management system consisted of production plan management subsystem, production scheduling management subsystem, storage and transportation management subsystem, quality management subsystem and production data management subsystem. The function orientation of the production Agent is to obtain tasks, organize the production process and coordinate the operation etc. [4,8]. In this case, the Agent works as a function one. And internally, it is a multi-agent system, consisting of planning Agent, scheduling Agent, transportation Agent and some other Agents.

The detailed structure of the Agent can be seen in Fig. 3. Interface Agent is designed to use the existing software system such as personnel and financial management information system. Through ORB or XML, the interface Agent can exchange information with the existing systems. Proxy Agent is responsible for the coordination of production Agent and other Agents in the management system. Knowledge Agent is responsible for the management of special knowledge in the production Agent, and offered information and knowledge to other function Agents. According to the actual situation of coal mine preparation production, four functional departments including planning department, scheduling department, transportation and marketing department and coal quality management department are assigned. Accordingly, four subfunction Agents are designed. The subfunction Agents themselves are multi-agent system, which needs to be further studied.

5. CONCLUSION

With the acceleration of information construction, the integrated manufacturing system of coal preparation plant develops rapidly. The effective integration of real-time data acquisition and the production management information in the system will be a good research topic for further study. In the present study, Agent is used as a tool and the integrated manufacturing system of coal mine preparation plant based on multi-agent is designed and studied with an application case. It reveals that such a system is helpful to the information production of coal mine preparation plant and provide a useful auxiliary decision support for the production command personnel.

Acknowledgements Financial support for this work, provided by the Natural Science Foundation of Jiangsu Province (BK2005022), is gratefully acknowledged.

REFERENCES

1. B.H. Li, The convention, standard and implementation guideline of CI M Systems, Beijing: The publish of weapon industry, 1994.
2. M.X. Lu, The production and management of coal preparation plant, Beijing: Coal Industry Press, 1991.
3. Y.L. Kuang and J.J. Deng, Application of Artificial Intelligence in Coal Preparation. Journal of China University of Mining & Technology. 30 (2001) 558–562.
4. J.J. Deng and Y.L. Kuang, Study on the Coal Preparation Plant Computer Integrated Manufacturing System Based on Multi-agent Technology. In: Proceedings of the 10th National Conference on Coal Preperation Science and Technology. XuZhou: China University of Mining and Technology Press, 2004.
5. Parunak. Multi-agent Systems-A Modem Approach to Distributed Artificial Intelligence. Massachusetts:The MIT Press, 2000.
6. M.M. Xia and H.C. Li, Intelligent Software Agent for Application in Distributed Information Management. Journal of Nanjing University of Aeronautics & A stronautics. 32 (2000) 591–596.
7. W. Shen, Agent-Based Systems for Intelligent Manufacturing. A State-of-the-Art. Knowledge and Information Systems. 1 (1991) 129- 156..
8. T. KTK, The realization of reference enterprise modeling architectures. Int. J. Computer Integrated Manufacturing. 12 (1999) 403-417.

2
CHAPTER

Theoretical Coupling Longitudinal-Transverse Model and Experimental Verification of Transverse Vibration of Rope for Multi-Rope Friction Hoisting System

Juan Wu[1,2] Ziming Kou[1,2]

[1] *Mechanical Engineering College, Taiyuan University of Technology, Taiyuan 030024, China*

[2] *Shanxi Provincial Engineering Laboratory for Mine Fluid Control, Taiyuan 030024, China*

ABSTRACT

Using the mass of time-varying length balance rope focused on the hoisting conveyance, the coupling longitudinal-transverse model of mine friction hoist was established by using of the Hamilton's principle. The modified Galerkin's method was used to discretize partial differential Eqs. The mine hoisting system was used to the example to analysis the relation between the load, velocity and transverse vibration of rope. The in situ tests were illustrated to evaluate the proposed mathematical model. The results showed that the modeling method can well represent the transverse vibration of rope.

Keywords: Hoisting system Vibration Rope Longitudinal-transverse Coupling

1 INTRODUCTION

Due to their ability to resist relatively large axial loads, ropes have been widely used in many different applications to support structures, conduct signals, and carry payloads. In this latter application, cables have played a vital role in the vertical transport systems such as elevators (Zhu and Ni 2000; Zhu et al.2001; Zhu and Chen 2005; Zhang 2007), cranes and particularly mine friction hoists (Kaczmarczyk and Ostachowicz 2003a; Sandilo and van Horssen 2014).

A common arrangement of mine friction hoist system is shown in Fig. 1. It comprises a driving friction wheel mounted in headgear, four or six hoist ropes, three balance ropes and two conveyances. In this system, the hoist ropes form vertical ropes connecting the friction wheel and conveyances. This configuration exhibits variable-length and transport speed during operation. It is classified in the category of one-dimensional continuous systems and consequently the displacement in situ depends on the time and a single spatial coordinate.

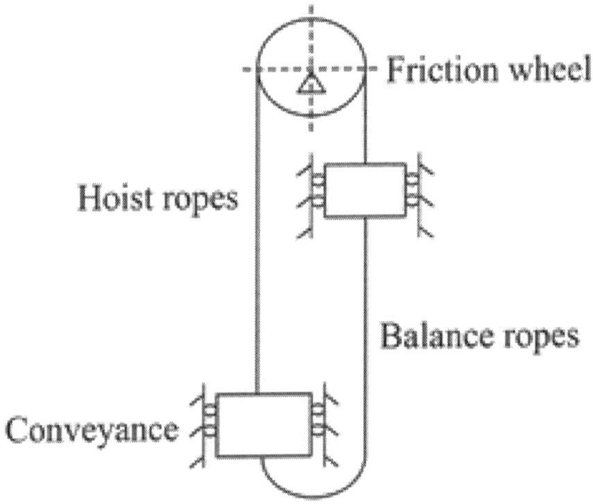

Figure 1. A friction hoisting installation

Elevator cables and friction hoist ropes are classified in the category of one-dimensional continuous systems. The vibrations of elevator ropes had been widely studied (Zhu and Ni 2000; Zhu et al. 2001; Zhu and Chen 2005; Zhang 2007; Sandilo and van Horssen 2014). Zhu and Ni (2000) generally investigated the stability characteristics of horizontally and vertically translating strings and beams with arbitrarily varying-length and with various boundary conditions. The forced response of translating media with variable length and tension was analyzed by Zhu and Chen (2005). An active control methodology using a point-wise control force and/or moment was developed by Zhu et al. to dissipate the vibratory energy of a translating medium with arbitrarily varying length. Zhang (2007) established the vibration governing Eq.s and energy Eq.s of coupled transverse-longitudinal vibration of elevator cables. Sandilo and van Horssen (2014) used a multiple-timescale perturbation method to construct formal asymptotic approximations of the solutions to show the complicated dynamical behavior of the string.

The longitudinal vibrations of ropes in mine hoist have been widely studied by many researchers. The single-rope mine drum is the catenary–vertical rope system and the coupled lateral–longitudinal dynamic response of the cables was studied by Kaczmarczyk and Ostachowicz (2003a, 2003b). Qin (2007) analyzed the longitudinal vibration of hoisting ropes by using the Lagrange principle. Li (2010) established Horizontal and vertical vibration models of the hoisting container under the steel guide and the vibration characteristic of hoisting container under different faults of steel guide was studied. To master the impact dynamical behaviors of the hoisting rope during the skip loading, the mechanical model of the coupled longitudinal-torsional vibration for hoisting rope were built by Cao (2010).

The importance of vibrations for ropes cannot be overestimated, especially in the context of the mining industry, where the safety of personnel and profitability rely on stable designs (Kaczmarczyk and Ostachowicz 2003). However; few transverse vibrations of ropes on coupling longitudinal-transverse model for mine hoist systems have been reported.

In this paper, the coupling longitudinal-transverse governing Eqs. of motion were established. The transverse dynamic responses of conveyance and ropes during descending cycle and ascending cycle were especially investigated. The in situ tests were performed to evaluate the proposed mathematical model.

2 THE GOVERNING EQUATIONS OF MOTION

The model was established on the following assumptions: (1) Young's modulus *E*, cross section area *A*, and mass per unit length were always constants; (2) The bending stiffness of rope and influence of air current were ignored; (3) The hoist ropes were simplified as single rope with the equivalent cross-section area. The rope was modeled as a taut string; (4) The mass of time-varying balance ropes was focused on conveyance.

A model of the mine hoist was shown in Fig. 2. The hoisting container was modeled as a rigid body of mass mm attached at the lower end x=l(t)x=l(t). The hoisting container was supported by a guide of stiffness kk and damping coefficient cc. The suspension of the container against the guide rails was assumed to be rigid. During its motion, the hoisting rope of mass per unit length ρρ had a variable length l(t)l(t) and an axial velocityV=l.(t)V=l.(t) where the over dot denoted time differentiation. The balance rope of mass per unit length ρ2ρ2 had a variable length H−l(t)H−l(t), where HH was the hoisting height.

The transverse displacement of the rope particle instantaneously located at spatial position xx at time tt was described by u(x,t)u(x,t) and longitudinal displacement was described by w(x,t)w(x,t). There was a boundary excitation e(t)e(t) in the upper end and the rigid roughness of the guide' surface was denoted by s(t)s(t).

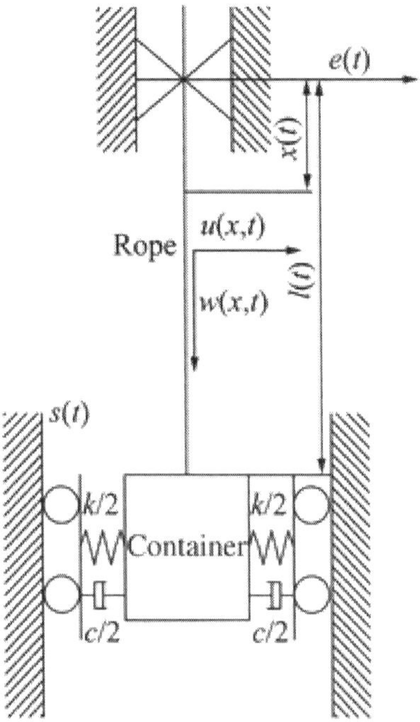

Figure 2. A model of mine hoists system

2.1 Free motion with damping

The Hamilton principle applied to the mine hoists yielded

$$\int_{t_1}^{t_2} (\delta T - \delta E_e - \delta E_g + \delta W)\mathrm{d}t = 0, \tag{1}$$

where T, E_e, E_g and W denoted the system kinetic energy, the rope elastic strain energy, the system gravitational potential energy and the damping force energy, respectively. The kinetic energy of the system was expressed as

$$T = \frac{1}{2} m_e V_c^2 + \frac{1}{2} \rho \int_0^l V^2 \mathrm{d}x, \tag{2}$$

where $m_e = m + (H - l(t))\rho_2$, V_c and V represented the total velocity at the conveyance and x on rope, respectively.

The elastic strain energy of the rope was

$$E_e = \frac{1}{2}k[u(l(t), t)]^2 + \int_0^l (P\varepsilon + \frac{1}{2}EA\varepsilon^2)\mathrm{d}x, \tag{3}$$

where EA represented the tensile stiffness of the rope, εε and P represented the strain measure in x on rope and the quasi-static tension of rope.

$$P = [m_e + \rho(l(t) - x)](g - a), \tag{4}$$

$$\varepsilon = w_x + \frac{1}{2}w_x^2 + \frac{1}{2}u_x^2, \tag{5}$$

where the subscript $()_x$ denoted the partial differentiation with respect to x, a and g represented the acceleration and gravitational acceleration

The gravitational potential energy of the cable was given by

$$E_g = E_{gs} - \int_0^l \rho g w(x, t)\mathrm{d}x - m_e g w_c(l, t), \tag{6}$$

where EgsEgs was the gravitational potential energy in the undeformed reference configuration.

The damping force energy was

$$W = \left(-c\frac{Dy(l(t), t)}{Dt} + ks + c\dot{s}\right) u(l(t), t). \tag{7}$$

And the boundary conditions were

$$w(0, t) = 0 \quad \delta w(0, t) = \delta w(x, t_1) = \delta w(x, t_2) = 0 \quad u(0, t) = 0 \tag{8}$$

The governing Eqs. for the system were yielded

$$\rho\left(w_{tt}+2vw_{xt}+\dot{v}\,w_x+v^2w_{xx}+\dot{v}\right)-P_x(1+w_x)-Pw_{xx}-\rho g$$
$$-EA\left(w_{xx}+3w_xw_{xx}+\frac{3}{2}w_x^2w_{xx}+u_xu_{xx}+w_xu_xu_{xx}+\frac{1}{2}u_x^2w_{xx}\right)=0,$$

(9a)

$$\rho(u_{tt}+2vu_{xt}+\dot{v}\,u_x+v^2u_{xx})-P_xu_x-Pu_{xx}$$
$$-EA(w_xu_{xx}+u_xw_{xx}+\frac{1}{2}w_x^2u_{xx}+w_xu_xw_{xx}+\frac{3}{2}u_x^2w_{xx})=0.$$

(9b)

The boundary conditions at x=l(t)x=l(t) were

$$m_e(w_{tt}+2vw_{xt}+\dot{v}+\dot{v}\,w_x+v^2w_{xx})+(P+EA\varepsilon)(1+w_x)-mg=0$$

(10a)

$$m_e\left(u_{tt}+2vu_{xt}+\dot{v}\,u_x+v^2u_{xx}\right)+\left(P+EA\varepsilon\right)u_x+c\left(u_t+vu_x\right)+ku-ks-c\dot{s}=0$$

(10b)

In this non-linear system, Eq. (9a) described the longitudinal dynamics of rope and Eq. (9b) represented the transverse dynamics of rope.

2.2 Boundary excitation

The cable cross-over motion at the friction wheel resulted in a boundary excitation. The boundary condition was formulated as

$$u(0,t)=e(t) \tag{11}$$

This boundary was inhomogeneous boundary conditions. The function u(x,t)u(x,t) with inhomogeneous boundary conditions was translated to the function y(x,t)y(x,t) with homogeneous boundary condition,

$$u(x,t)=y(x,t)+h(x,t), \tag{12}$$

where $h(x,t) = a_0(t) + a_1(t)\frac{x - x_0}{x_1 - x_0}$.

Thus, when the excitation terms were taken into consideration, the response of the system was governed by

$$\rho\left(w_{tt} + 2vw_{xt} + \dot{v}\,w_x + v^2 w_{xx} + \dot{v}\right) - P_x(1 + w_x) - Pw_{xx}$$
$$-EA(w_{xx} + 3w_x w_{xx} + \frac{3}{2}w_x^2 w_{xx} + (y_x + h_x)(y_{xx} + h_{xx}) + w_x(y_x + h_x)(y_{xx} + h_{xx}) + \frac{1}{2}(y_x + h_x)^2 w_{xx}) - \rho g = 0,$$

(13a)

$$\rho\left(y_{tt} + 2vy_{xt} + \dot{v}\,y_x + v^2 y_{xx}\right) - P_x y_x - Py_{xx}$$
$$-EA\left(w_x y_{xx} + y_x w_{xx} + \frac{1}{2}w_x^2 y_{xx} + w_x y_x w_{xx} + \frac{3}{2}(y_x + h_x)^2(y_{xx} + h_{xx})\right) = f(x,t).$$

(13b)

where
$f(x,t) = P_x h_x + Ph_{xx} + EA(w_x h_{xx} + h_x w_{xx} + \frac{1}{2}w_x^2 h_{xx} + w_x h_x w_{xx}) - \rho(h_{tt} + ah_x + 2vh_{xt} + v^2 h_{xx})$.

The boundary conditions at x=l(t)x=l(t) were the same with Eqs. (9a) and (9b).

3 NUMERICAL SOLUTIONS

3.1 Spatial discretization

The modified Galerkin's method was employed to discretize the governing partial differential equations. The instantaneous orthonormal eigenfunction with time-varying length l(t)l(t) were used as the trial functions. They can be expressed as $\varphi_j(x,t) = \frac{1}{\sqrt{l(t)}}\varphi_j(\xi)(j = 1,2,3\ldots)$, $\xi = \frac{x}{l(t)}$, $\varphi_j(\xi) = \sqrt{2}\sin(\frac{2j-1}{2}\pi\xi)$.

The solutions of Eq. (13) were assumed in the form

$$w(x,t) = \sum_{j=1}^{n} p_j(t)\varphi_j(x,t) = \frac{1}{\sqrt{l(t)}}\sum_{j=1}^{n} p_j(t)\varphi_j(\xi), \tag{14a}$$

$$y(x,t) = \sum_{j=1}^{n} q_j(t)\varphi_j(x,t) = \frac{1}{\sqrt{l(t)}}\sum_{j=1}^{n} q_j(t)\varphi_j(\xi). \tag{14b}$$

where pj(t)pj(t) and qj(t)qj(t) were the generalized coordinates, and n was the number of included modes. Substituting Eq. (14) into the governing Eq. (13) and boundary conditions of Eq. (9), multiplying the governing equation by $\frac{1}{\sqrt{l(t)}}\varphi_j(\xi)$, integrating it from x=0x=0 to $l(t)$, yielded the discretized equations of motion.

$$\begin{bmatrix} M_1 & 0 \\ 0 & M_2 \end{bmatrix}\begin{bmatrix} P'' \\ Q'' \end{bmatrix}+\begin{bmatrix} C_1 & 0 \\ 0 & C_2 \end{bmatrix}\begin{bmatrix} P' \\ Q' \end{bmatrix}+\begin{bmatrix} K_1 & 0 \\ 0 & K_2 \end{bmatrix}\begin{bmatrix} P \\ Q \end{bmatrix}=\begin{bmatrix} CP_1 \\ CP_2 \end{bmatrix}+\begin{bmatrix} F_1 \\ F_2 \end{bmatrix} \quad (15)$$

where P=[p1,p2,...,pn]TP=[p1,p2,...,pn]T and Q=[q1,q2,...,qn]TQ=[q1, q2,...,qn]T were the vectors of generalized coordinates, *M,C, K, CP* and *F* were symmetric mass, damping, stiffness matrices, coupling and generalized force matrices. Entries of these matrices were given in Appendix 1.

3.2 Example and discussions

The example of mine hoisting system was illustrated to evaluate the proposed mathematical model. The maximum hoisting height and acceleration were 263 m and 0.6 m/s^2. The mass of container was 46000 kg. There were six hoisting ropes and mass per unit length was 8.52 kg/m. And the mass per unit length of three balance ropes was 16.8 kg/m. The normal load range was 0, 7, 14, 21, 28, 35, 42 and 49 kN and the speed was 6, 7 and 8 m/s.

The prescribed displacement, velocity, acceleration curves were utilized as the input of Eq. (15) to obtain the longitudinal and transverse vibration response of rope and conveyance. The transverse vibration responses of rope were specifically analyzed in this paper. The boundary excitation e(t)e(t) was represented by

$$e(t) = 0.001 \times \sin(1.45\pi t) \quad (16)$$

The s(t)s(t) was given by

$$s(t) = 0.001 \times \sin(0.6\pi t). \quad (17)$$

The transverse dynamic responses of rope at (F1) and (F8) were shown in Figs. 3 and 4. The vibration of rope in descending cycle was different from that in ascending cycle. In descending cycle, the amplitude of transverse dynamic responses was gradually increased with increasing the length of rope. In ascending cycle, the amplitude of transverse dynamic responses was gradually increased with decreasing the length of rope and the frequency of transverse dynamic responses was higher near to the pithead.

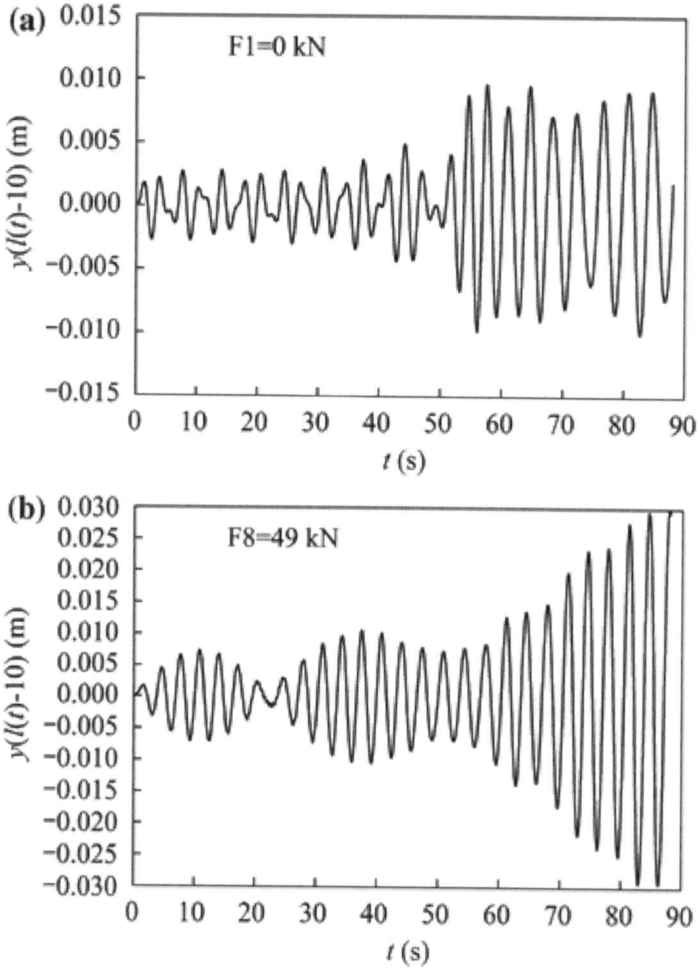

Figure 3. The transverse response time histories of rope at 6 m/s in descending cycle.

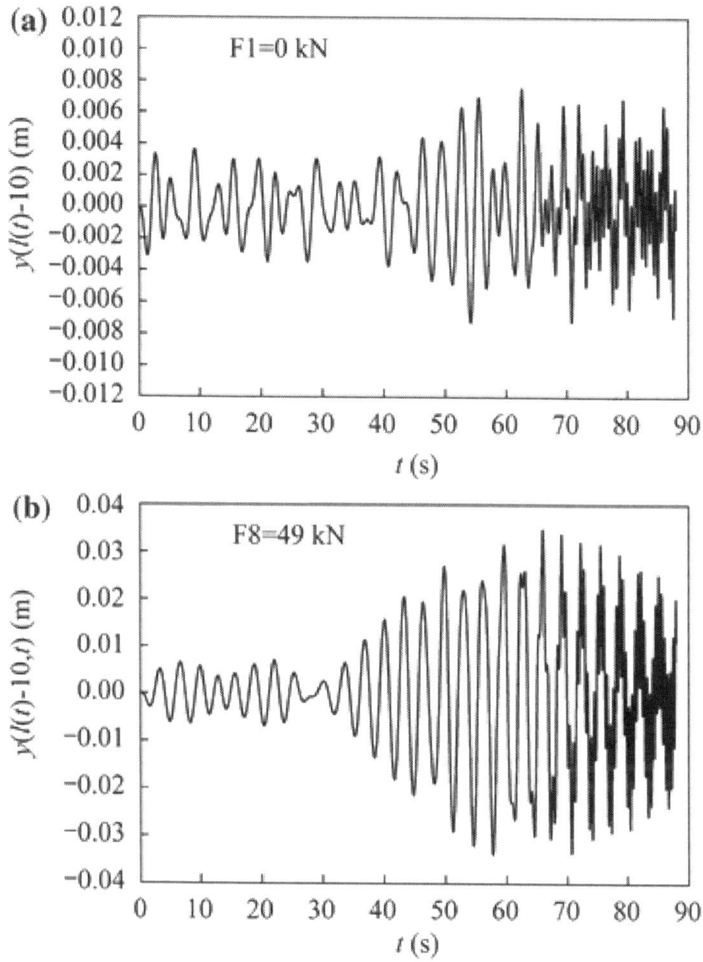

Figure 4. The transverse response time histories of rope at 6 m/s in ascending cycle.

Figure 5 showed the variations of maximum amplitude at different loads and velocities. Generally speaking, the maximum amplitudes in ascending and descending cycle showed the similar trend and increase with the increase of load. The maximum amplitudes in ascending cycle were the higher and those in descending cycle were lower. The maximum amplitudes showed a fluctuation with the increase of speed and the maximum amplitudes at 8 m/s were lower than those at 6 and 7 m/s.

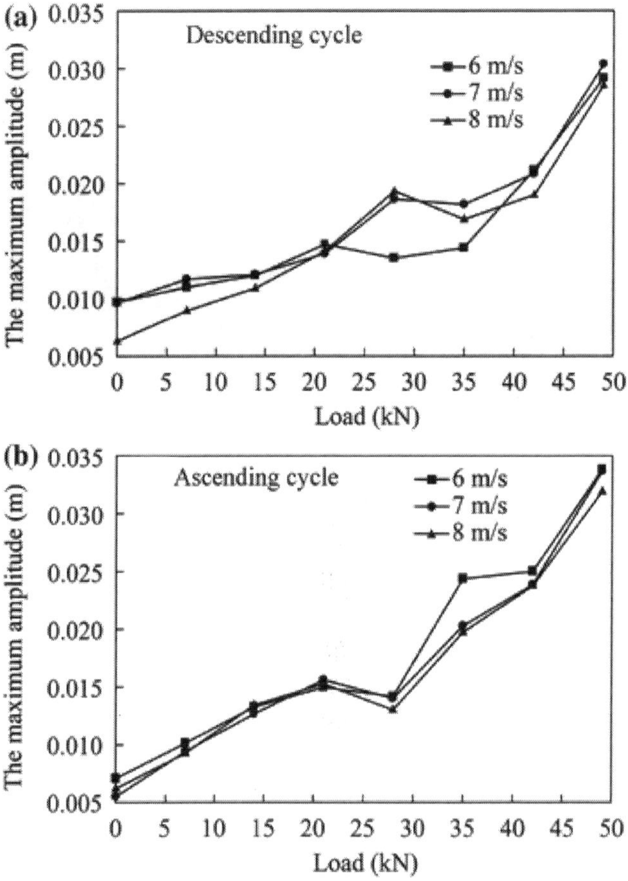

Figure 5. Variations of maximum amplitudes of rope at different loads and velocities: **a** in descending cycle and **b** in ascending cycle.

4 IN-SITU TESTS

Four INV9822 acceleration sensors were installed on the four ropes respectively with a distance of 10 m from the conveyance, shown in Fig. 6. The data acquisition analyzer was analyzer INV3060S and the data acquisition and signal analysis platform was DASP-V10, which were made by the Dongfang Vibration and Noise Technology Institute, Beijing, China.

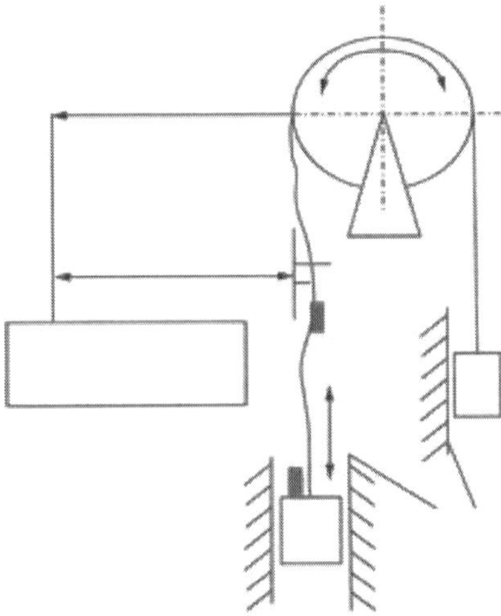

(a) Overall test set-up

(b) Sensors on ropes

Figuer 6. Test setup

The history of response time of the rope for in situ tests was shown in Figs. 7, 8 and 9 was the amplitude-frequency characteristics of transverse vibration. Though the influence of complicated conditions usually occurred in in situ tests, boundary excitations in the upper end and the rigid roughness of the guide' surface would be the major factors. Compared the response time histories of rope and amplitude-frequency characteristic obtained by the present analysis with those obtained experimentally, the mathematical model and numerical solution were in an acceptable level of accuracy as compared with experimental data. It suggests that the models were validated.

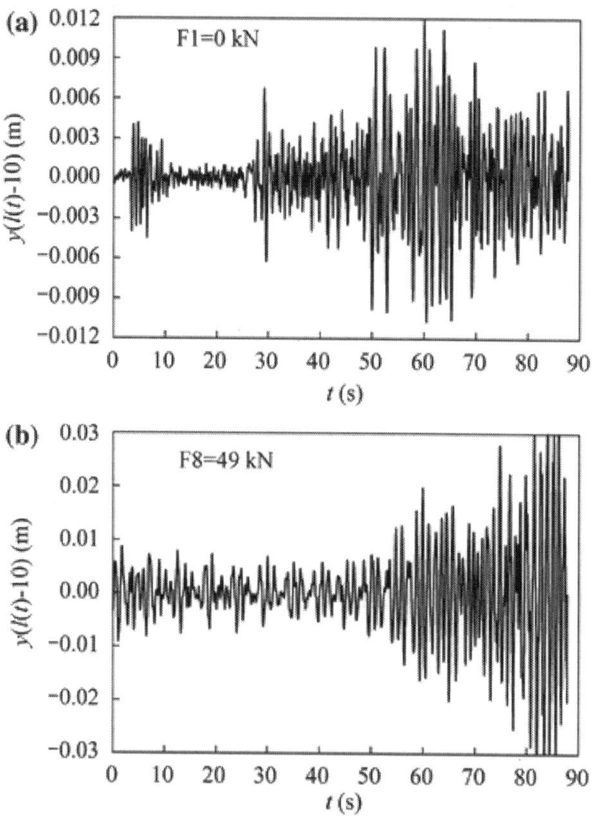

Figure 7. The transverse response time histories of rope in descending cycle.

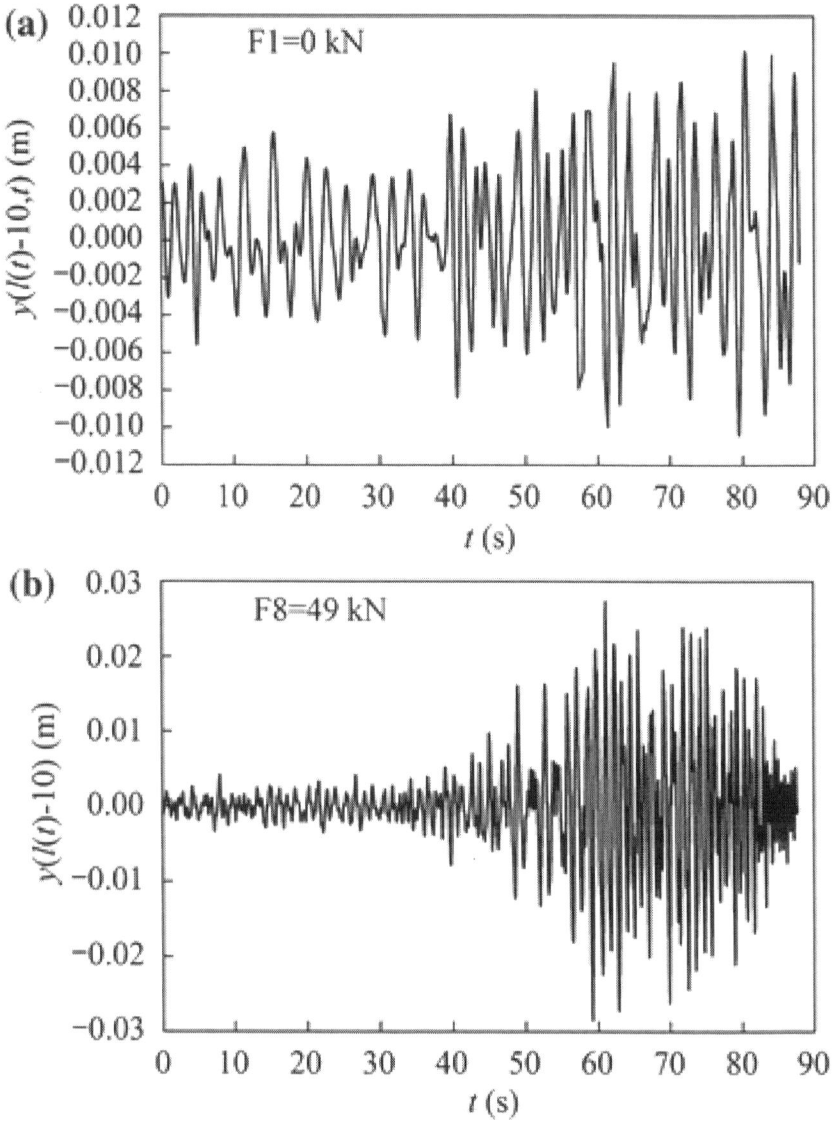

Figure 8. The transverse response time histories of rope in ascending cycle

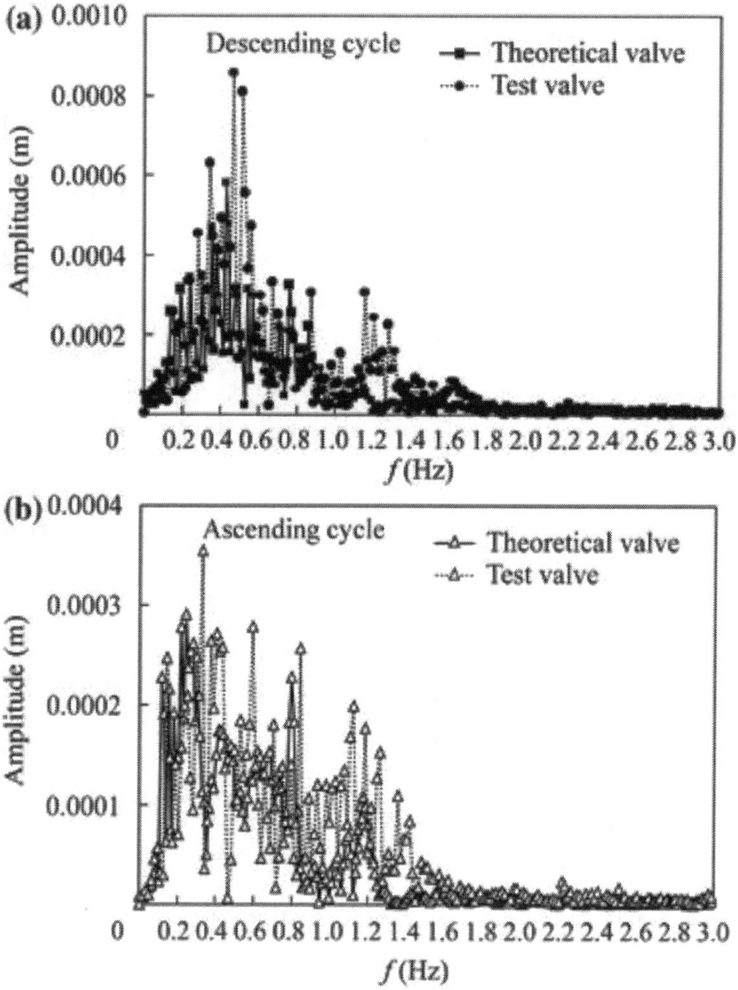

Figure 9. The amplitude-frequency characteristic of transverse vibration

5 CONCLUSIONS

1. The Hamilton's principle was employed to study the coupling longitudinal-transverse governing equations of motion for the rope in mine hoisting system. The partial differential equations of motion of this system have been derived.

2. The transverse vibration of rope had been theoretically studied. The transverse vibration of rope in ascending cycle was stronger than that in descending cycle. The maximum amplitudes in ascending and descending cycle showed the similar trend and increase with the increase of load. The maximum amplitudes at 8 m/s were the lowest.

3. The result obtained by the mathematical model was compared with result obtained experimentally. The present model was proved to be accurate sufficiently by compared with test results.

4. The theoretical formulation and numerical solution scheme provided an efficient method for the analysis of the transverse vibration of the rope for mine hoist system.

ACKNOWLEDGMENTS

The authors are grateful to the International S & T Cooperation Program of China (2011DFA72120) and NSFC (51205272) for financial support.

REFERENCES

1. Cao GH (2009) Impact dynamic behaviors of mine hoisting rope during loading process. China University of Mining & Technology, Xuzhou (in Chinese with English Abstract)

2. Kaczmarczyk S, Ostachowicz W (2003a) Transient vibration phenomena in deep mine hoisting cables. Part 1: mathematical mode. J Sound Vib 262:219–244

3. Kaczmarczyk S, Ostachowicz W (2003b) Transient vibration phenomena in deep mine hoisting cables. Part 2: numerical simulation of the dynamic response. J Sound Vib 262:245–289

4. Li ZF (2010) Study on the vibration characteristic and typical fault diagnosis of mine hoist system. China University of Mining & Technology, Xuzhou (in Chinese with English Abstract)

5. Qin D (2007) Study on safety of coal mine vertical friction hoisting system based on dynamic. Hefei University of Technology, Hefei (in Chinese with English Abstract)

6. Sandilo SH, van Horssen WT (2014) On variable length induced vibrations of a vertical string. J Sound Vib 333:2432–2449

7. Zhang P (2007) Theoretic and test research on dynamic behaviors of high speed elevator suspended system. Doctor's thesis, Shanghai Jiao Tong University (in Chinese with English Abstract)

8. Zhu WD, Chen Y (2005) Forced response of translating media with variable length and tension: application to high-speed elevators. Proc Inst Mech Eng, Part K 219(1):35–53

9. Zhu WD, Ni J (2000) Energetics and stability of translating media with an arbitrary varying length. J Vib Acoust, Trans ASME 122(3):295–304

10. Zhu WD, Ni J, Huang J (2001) Active control of translating media with arbitrarily varying length. J Vib Acoust, Trans ASME 123(3):347–358

3

CHAPTER

Microstructures and Mechanical Properties of a Martensitic Steel Welded with Flux-Cored Wires

Weiqiang Zhang[1] ,Zhimin Zhong[1] , Shaoguang Kang[2]

[1]*School of Environmental and Chemical Engineering, Shenyang Ligong University, Shenyang 110159, China*

[2] *China Coal Zhangjiakou Coal Mining Machinery Co. Ltd., Zhangjiakou 076250, China*

ABSTRACT

This study was focused on wear-resistant material prepared by CO_2 GMAW method to basically determine the alloys and hardfacing technology which could be employed into mining equipment. Three flux-cored wires with different chemical compositions, marked by DM-I, DM-II and DM-III were used in welding a martensitic steel for the comparison of the microstructures and mechanical properties of the welding metals and HAZ. The results show that DM-I weld metal contains martensite plus other reinforced phases, while both DM-II and DM-III alloys lead to mainly martensitic microstructures with little precipitates. Among all of the tested materials the DM-I alloy has the highest hardness and wear resistance but the lowest impact toughness among the experimental weld metals. Meanwhile, the hardness, impact toughness and wear resistance of the DM-II and DM-III alloys are close to those of the base metal but the DM-III is not as strong as DM-II. Dimples can be found on the impact fracture of the weld metals, but the steel fracture consists of both cleavage and dimples. Thus HAZ has less risk for cracking than the welds, and the

DM-II wire which produces the weld having identical mechanical properties to the base metal is suitable for further application.

Keywords: Hardfacing Microstructure Mechanical property Flux-cored wire GMAW

1 INTRODUCTION

Wear of materials occurs frequently in coal mining industries. To solve this problem, numerous hardened metals and advanced technologies are used to resist the wear from hard substances such as coals, rocks or ores. Surfacing with hard filler metals or alloys is one of the most effective methods for maintenance of the worn materials, permitting prolonged equipment lives. By now, remarkable progress has been made in developing efficient and flexible surfacing technology to improve both wear and corrosion resistance of materials, and lead to wide applications in industry (Menon 1996; Prasad et al. 1998). However, for those martensite-hardened steels, filling hard metals with different compositions is considered as a great difficulty due to a strong tendency of cracking. To obtain a crack-free hard surface both chemical composition and hardfacing technology need to be carefully designed.

Although the influences of chemical compositions of filler metals on the microstructure of the deposit layers are important to determine the mechanical property and performances, much attention has also been paid to develop filler metals to improve the compatibility to the base materials, which might be one of the main problems in both surfacing and laser cladding, and requires the suitably designed composition of the coating alloys (Zollinger et al. 1998). Examples lie in the works on the influence of Cr, Nb and C on the microstructure as well as hardness and wear resistance of deposit layers on nickel-based alloy (Su and Chen 1997; Kesavan and Kamaraj 2010). Other authors promoted different iron-based crack-resisting filler alloys, showing potentially advantageous performances of the hardened surfaces, among which carbide-forming elements are mostly considered to use for high hardness of the workpieces (Ingis et al. 1992; Wang et al. 2004; Yang et al. 2011, 2013; Zhang and Ren 2004). In recent years, rare earth elements are found in refining microstructures of hardfacing metals (Hao et al. 2011a,b). For those hardened surfaces, different alloy systems lead to a great diversity in wear behavior and performances, mainly depending on the microstructures (Wu and Wu 1996; Agarwal and Dahotre 1998; Heath and Skora 1998; Kim and Kim 1999; Przybylowicz and Kusinski 2000).

For an iron-based alloy system, element C, Cr, Mo or V are usually added to increase the hardness and wear resistance of the coatings, which also leads to great changes in other performances. Especially for a martensitic base metal, surfacing with hard alloys has a high risk of cracking. Therefore the microstructures and mechanical properties of a hardened base metal surfaced with different alloys remain much interesting for scientists and engineers. Herein, three flux-cored weld wires are used in surfacing a hardened steel by comparison of the microstructures and mechanical properties to obtain further understanding of the effects of alloying elements and expand industrial applications of surfacing technology.

2 EXPERIMENTAL MATERIALS AND METHODS

A hot-rolled low-carbon martensitic steel in delivery state was employed as base metal with chemical composition listed in Table 1 and the Rockwell hardness was around 40HRC. The steel plate was machined into specimens of the dimension 1600 mm × 160 mm × 40 mm with the Y-type grooves made by the angle of 90°. Then CO_2 GMAW method was used in welding these specimens. The technological parameters were chosen as the output secondary voltage and electric current are, respectively, 30 V and 200 A; the gas flux 7.0 mL per second, and the welding speed was controlled by 3.0 mm/s. Three flux-cored welding wires with the diameter 1.2 mm were chosen as the filling metals, and the compositions of the deposited iron-based alloys numbered with DM-I, DM-II and DM-III, are shown in Table 2. The samples were carefully processed to ensure that microstructure observation and mechanical property test can be performed exactly within the corresponding zones. Optical microscope and Shimadzu SSX-550 SEM were used for the microstructure observation of the weld metals and HAZ, and thin foil specimens of the weld were prepared and observed on TEM. The phases contained in the weld metals were also analyzed by XRD method.

Table 1. Chemical composition of the tested steel (wt%)

C	Si	Mn	Cr	P	S	Ti	B
0.17	0.30	1.45	0.19	0.015	0.006	≤0.02	≤0.0018

Table 2. Chemical composition of deposited metals (wt%)

Element	C	Mn	Si	Cr	Mo	V	B	P	S	Fe
DM-I	0.470	0	0	5.000	1.230	0.120	0.351	0.010	0.014	Balance
DM-II	0.120	1.352	0.570	3.200	0.980	0	0	0.015	0.012	Balance
DM-III	0.110	1.800	0.610	1.960	0.490	0	0	0.016	0.005	Balance

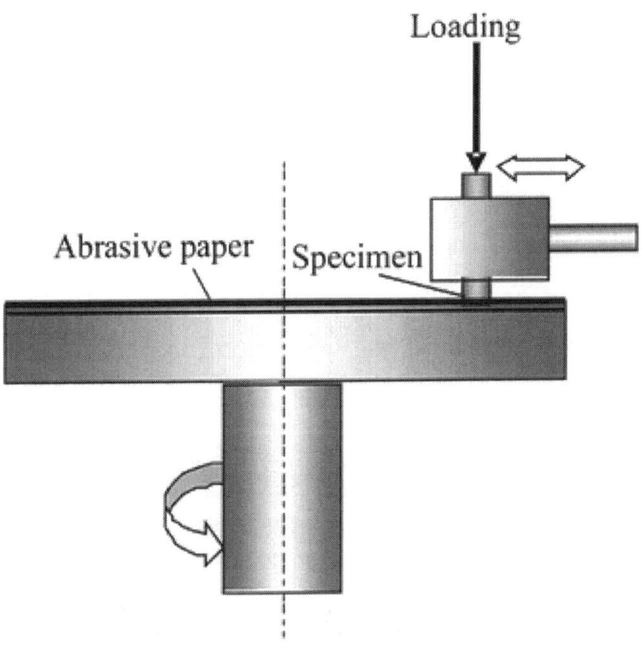

Figure 1. Schematic diagram of the wear test

Rockwell hardness was tested from the centerline of the weld towards the base metal. The impact toughness of each zone is tested at ambient temperature by using the specimens with U notches prepared from the welded steel. Wear experiment which is schematically shown in Fig. 1, was carried out by using the cylinder specimens with the dimension $\phi 5$ mm × 10 mm moving on the abrasive papers of 400 mesh for different distances, and then mass losses of these specimens were measured.

3 RESULTS AND DISCUSSION

3.1 Microstructures

Figure 2 shows the microstructures of the martensitic steel and the weld metals, respectively, with the wires DM-I, DM-II and DM-III. The initial microstructure of the steel is mainly low-carbon martensite as indicated in Fig. 2a. The weld metals are distinguished from the base metal in microstructures due to different chemical compositions. Figure 2b shows DM-I weld microstructure, in which carbides form on the grain boundaries, and precipitation also occurs within the grains. Addition of relatively much alloying elements, Cr, Mo and B makes the deposited alloy DM-I have proeutectoid composition, which leads to the typical solidified microstructure with equiaxed grains surrounded by carbides or other secondary phases. Both of Fig. 2c, d show similar morphologies of the martensite microstructures, except that the microstructure in Fig. 2d is finer than Fig. 2c. The DM-II alloy contains similar percentages of C, Si and Mn to those of DM-III but more contents of Cr and Mo than DM-III, which causes more precipitates in DM-II alloy. The phase constitutions in the three alloys can be qualitatively characterized by XRD method. As shown in Fig. 3, the results indicate that the alloy DM-I produces not only the hardening martensite, but also $M_{23}C_6$-type carbides and boron compound, which contributes greatly to high hardness of the weld, whereas the DM-II and DM-III weld metals contain almost identical phases with little carbides or other secondary compounds found.

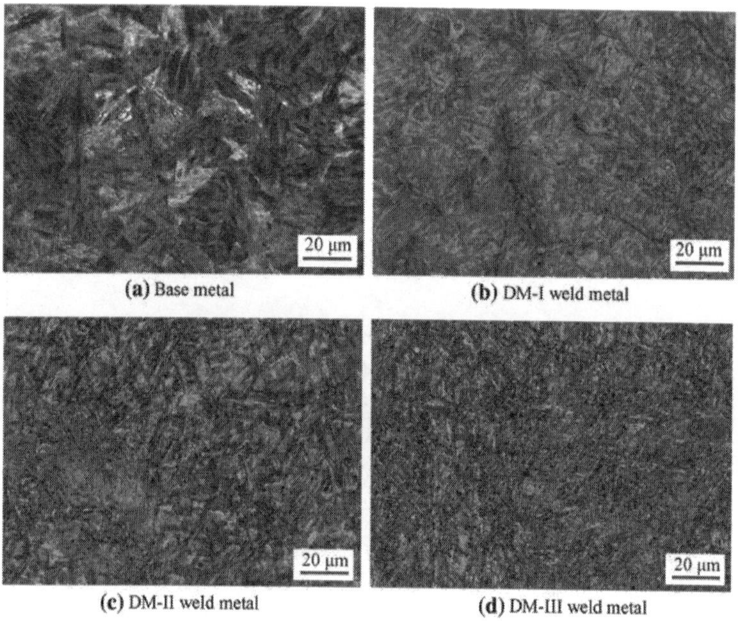

Figure 2. Microstructures of the weld metals and the base metal

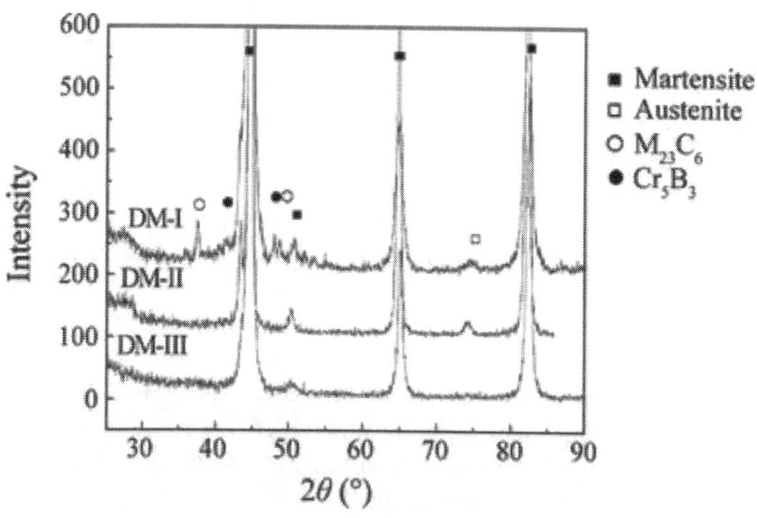

Figure 3. XRD patterns for the weld metals with DM-I, II and III flux-cored welding wires

By TEM method, the microstructures of the DM-I and DM-II weld metals are compared in Fig. 4. For the DM-I weld, martensite with fine sub-grains and twins can be observed concurrently with carbides as shown in Fig. 4a, whereas the microstrucutres for the DM-II weld and the base metal is martensite that mainly consists of dislocation.

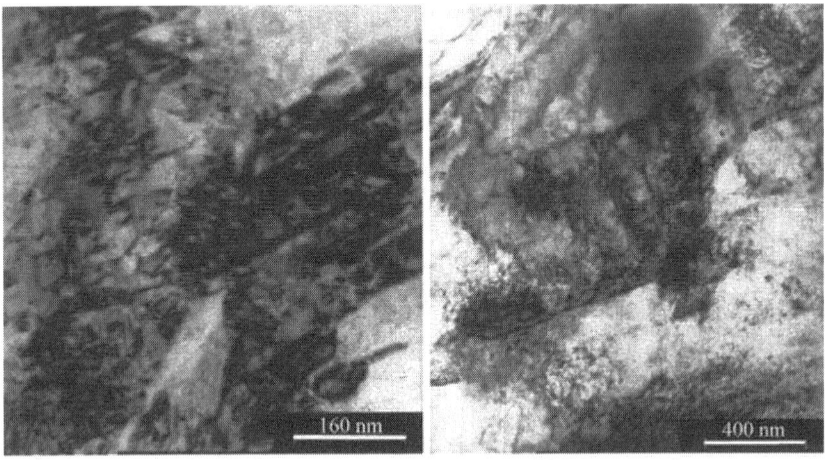

(a) DM-I weld metal (b) DM-II weld metal

Figure 4. TEM images of the microstructures of weld metals showing precipitates and dislocations in the subgrains

The microstructures of HAZ is shown in Fig. 5. Observation on the microstructures of HAZ shows different characteristics due to the highest temperatures locally reached during welding. Welding bond that is adjacent to the weld differs from the weld metal where solidification occurs from the base metal where only solid transformations take place. Martensitic transformation occurs in the superheated zone and completely quenched zone since both of the zones are completely austenized and fast cooled, which are easily distinguished by the grain size. The completely quenched zones consists of finer martensite than the superheated zone due to the temperature that is slightly higher than A_{c3}. There is a narrow fine-grained zone adjacent to the completely quenched zone where the steel is locally heated up to the temperature range between A_{c1} and

A_{c3} and then quenched, so that dual-phased microstructure (ferrite + martensite) can be observed, which is indicated as the subcritically quenched zone. As the temperature is lower than A_{c1}, the initial martensitic structure is somewhat tempered.

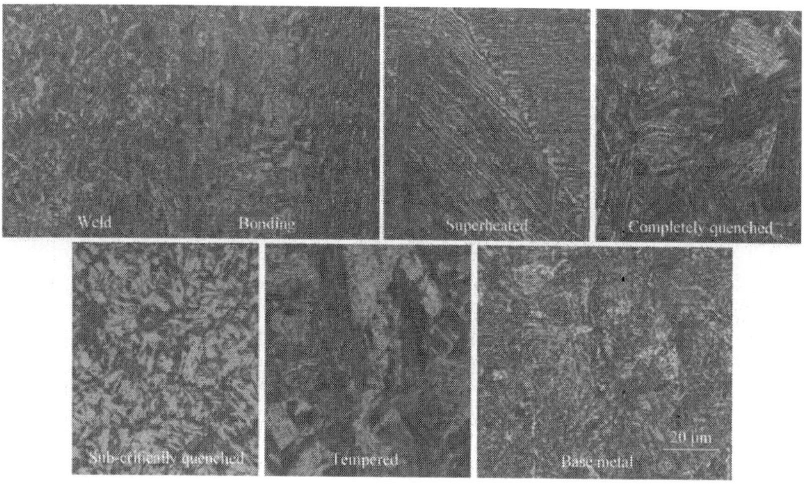

Figure 5. The microstructures of HAZ for the steel welded by DM-II wire

3.2 Mechanical properties

Figure 6 indicates the hardness and impact toughness distribution along with the weld metals and HAZ. The weld deposited with DM-I alloy shows the highest hardness among the three depositing metals, whereas the DM-II weld which is nearly identically hard to the base metal, is approximately 5HRC higher than DM-III in hardness. Then the DM-II alloy is suitable for the requirement of the same hardness for the weld and base metal. In the HAZ, there is a softened zone located 7–10 mm away from the central line of the welds. This softening effect is resulted from the sub-critical temperature reached in the welding so that the dual phase microstructure is formed.

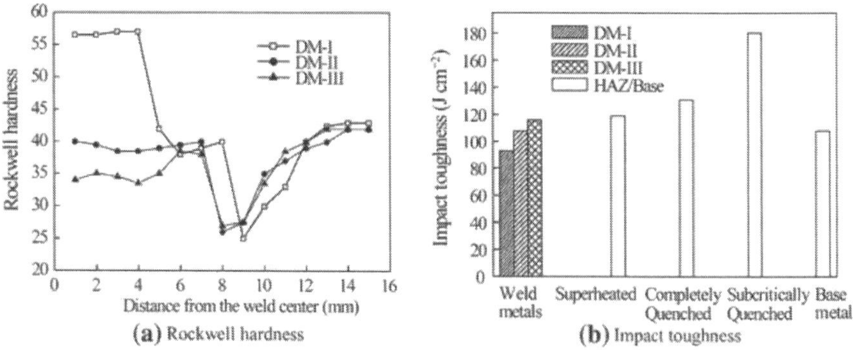

Figu 6. Mechanical properties of the weld metals and HAZ

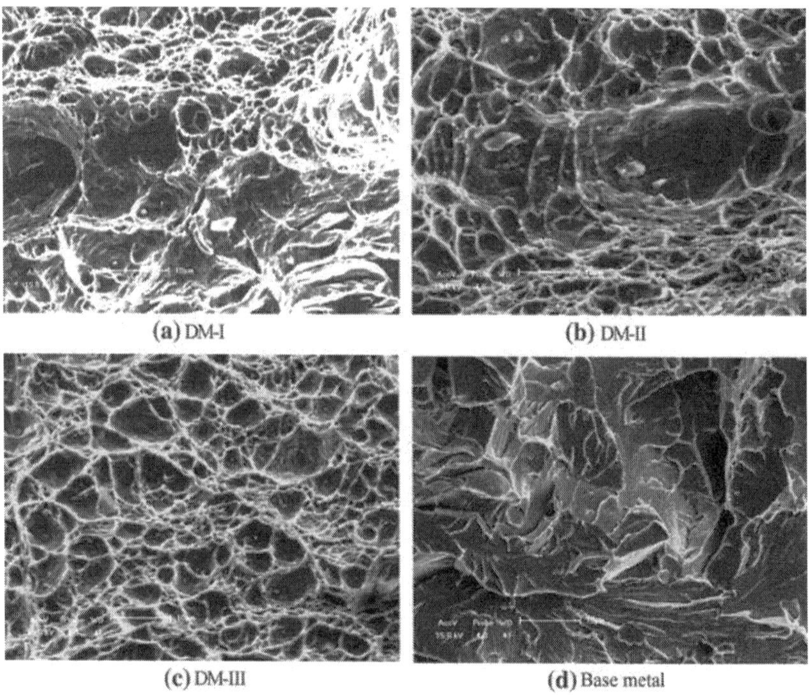

Figure 7. Fractography for impact samples of the weld metals and base metal

The impact toughness of the weld metal is dependent on the local hardness. The DM-I weld which is the hardest has the lowest toughness among the three welds, and obviously lower than that of

HAZ and the base metal. However, the highest toughness appears at the incompletely-quenched zone where the formation of two phases, ferrite + martensite, increases the resistance to crack propagation. Although the superheated zone has coarsened grains, the impact toughness is a slightly higher than the base metal.

The fracture observed via SEM on the impact samples of the three weld metals and the base metal are shown in Fig. 7. For DM-I weld as shown in Fig. 7a, crack propagates in a combined mechanism with inclusions observed, indicating mixture of dimples and cleavage characteristics. Dimples are mainly formed on the fracture of both DM-II and DM-III welds, except that the dimples are more uniform and smaller in DM-III weld. The experimental steel that did not experience a heat cycling during welding is not strong in resisting crack propagation.

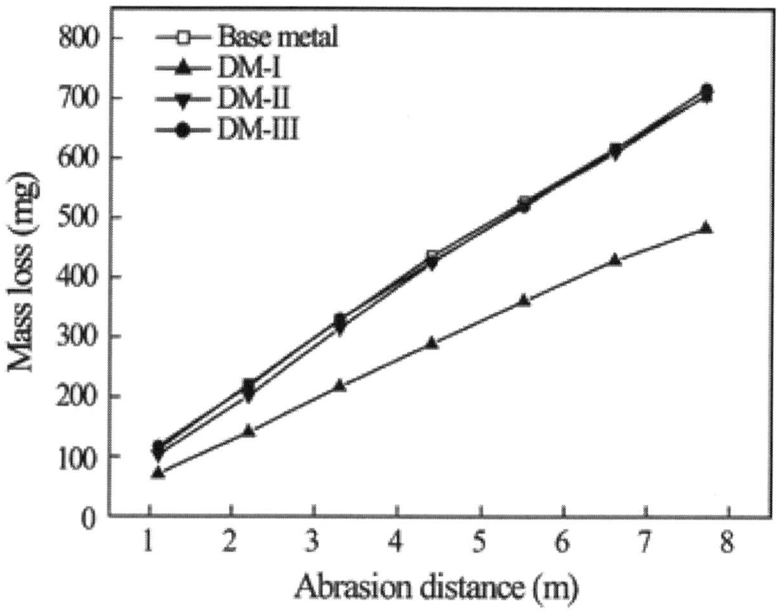

Figure 8. Comparison of wear resistance of weld metals and base metal

Figure 8 shows the comparison of wear resistance of the three depositing metals and the base metal with respect to the abrasion distance. The hardest DM-I alloy display the lowest mass loss, i.e. the highest wear resistance, but the mass losses of DM-II and DM-III alloys are identical to that of the base metal, although the hardness of DM-III alloys is lower. Wear resistance correlates with the hardness as well as other properties such as toughness due to the fact that wear consists of micro-fracture depending upon the toughness of the materials. Thus the increment in the impact toughness of DM-III compensates for the decrease in wear resistance resulted from the reduction of hardness.

4 CONCLUSIONS

*(1)*The martensitic steel welded by DM-I weld metal contains reinforced phases, while both DM-II and DM-III alloys have little precipitates but mainly low carbon martensite formed during welding. In HAZ of the steel, different microstructure with martensite forms in welding, with coarse grains observed in the superheated zone. Compared with the weld and base metal, low hardness appears in the subcritically-quenched zone of the HAZ, but the impact toughness is higher than the base metal, meaning HAZ has weak cracking tendency during welding.

*(2)*Compared with the initial state of the steel, the DM-I weld metal has the highest hardness and wear resistance but the lowest impact toughness, while the DM-III has somehow lowest hardness although the wear resistance is almost identical to the base metal. However, the mechanical properties of DM-II alloy are close to those of the base metal, showing a proper weldability as well as applicability for such cases.

*(3)*Dimples are mainly found on the impact fracture of the weld metals, but impact fracture of both HAZ and base metal consists of both cleavage and dimples.

REFERENCES

1. Agarwal A, Dahotre NB (1998) Pulse electrode deposition of superhard boride coatings on ferrous alloy. Surf Coat Technol 106(2–3):242–250

2. Hao F, Liao B, Li D, Liu L, Dan T, Ren X, Yang Q (2011a) Effects of rare earth oxide on hardfacing metal microstructure of medium carbon steel and its refinement mechanism. J Rare Earths 29(6):609–613

3. Hao F, Li D, Dan T, Ren X, Liao B, Yang Q (2011b) Effect of rare earth oxides on the morphology of carbides in hardfacing metal of high chromium cast iron. J Rare Earths 29(2):168–172

4. Heath G, Skora J (1998) New materials and process developments for preventative maintenance and repair in cement plants. Cem World 2:359–365

5. Ingis I, Murphy EV, Ocken H (1992) Performance of wear-resistant iron base hardfacing alloys in valves operating under prototypical pressurized water reactor conditions. Surf Coat Technol 53(1):101–106

6. Kesavan D, Kamaraj M (2010) The microstructure and high temperature wear performance of a nickel base hardfaced coating. Surf Coat Technol 204(24):4034–4043

7. Kim HJ, Kim YJ (1999) Wear and corrosion resistance of PTA weld surfaced Ni and Co based alloy layers. Surf Eng 15(6):495–501

8. Menon R (1996) New developments in hardfacing alloy. Weld J 75(2):43–49

9. Prasad BK, Dasgupta R, Jha AK, Modi OP, Das S, Yegneshwaran AH (1998) Hardfacing: a technique for combating abrasive/erosive wear. Mater Trans JIM 39(12):1191–1196

10. Przybylowicz J, Kusinski J (2000) Laser cladding and erosive wear of Co-Cr-Mo-Si coating. Surf Coat Technol 125(1–3):13–18

11. Su YL, Chen KY (1997) Effect of alloy additions on wear resistance of nickel based hardfacing. Weld J 76(3):143–150

12. Wang X, Song S, Zou Z, Qu S (2004) Development of new type of wear and crack resistant hardfacing electrode. Trans Nonferr Metals Soc China 14(4):660–664

13. Wu W, Wu L (1996) The wear of behavior between hardfacing materials. Metall Mater Trans A 27(1):3639–3648

14. Yang K, Zhang Z, Hu W, Bao Y, Jiang Y (2011) A new type of submerged-arc flux-cored wire used for hardfacing continuous casting roll. J Iron Steel Int 18(11):74–79

15. Yang K, Yang Q, Bao Y (2013) Formation of carbonitride precipitates in hardfacing alloy with niobium addition. Rare Met 32(1):52–56

16. Zhang Y, Ren D (2004) Effects of strong carbide forming elements in hardfacing weld metal. J Univ Sci Technol Beijing 11(1):71–75

17. Zollinger OO, Beckham JE, Monroe C (1998) What to know before selecting hardfacing electrodes. Weld J 77(2):39–43

4

CHAPTER

Functional Simulation of the Method for the Coal Preparation for Combustion in the Thermal Power Plant

Natalia Vasilievna Fedorova, Vasily Alexandrovich Mokhov and Elena Alexandrovna Krivobok

Platov South-Russian State Polytechnic University (NPI)

ABSTRACT

Russia's energy policy for the coming years provides for greater use of coal fuel, combined with the introduction of non-waste technologies of its combustion. Russia has taken to burn coal with ash content of various indicators, including those of high. The question of efficient use of coal in thermal power plants (TPP) of Russia by his deep preparation for burning is considered in this article. One way to solve of this problem is the construction of modern coal preparation factory directly or near the power plant. The criteria of efficiency of coal combustion are systematized. The modern methods of mineral raw materials enrichment used in the mining, metallurgical and chemical industries are considered. The parameters of coal preparation equipment offered in the Russian market by manufacturers from different countries are investigated. The systematic analysis of production processes in the TPP by functional modeling techniques based on IDEF0 diagrams and hierarchies analysis of T. Saaty is performed, allowing to prove the correctness of the proposed solution of the problem of coal enrichment in TPP prior to combustion and to provide an option of the technological chain of operations performed. The technical and economic effect after object modernization is defined. Coal deep preparation for burning, including the introduction of coal preparation factory on TPP, will create the environmentally friendly technology of fuel and energy coal-fired complex.

Keywords: high ash content coal; coal preparation; activation of combustion; analysis of technological processes; systems analysis; process modeling; rating

INTRODUCTION

Russia's energy policy [1] provides for greater use of coal fuel, combined with the introduction of non-waste technologies of its combustion.

According to the Russian National Standards, the ash-content of coking coal shall not exceed 5-6%, and the ash-content of the steam coal shall amount to 13-18% [2-3]. The reduction of the ash content of the coal (coal preparation), supplied for the burning, should be based on the feasibility study and be profitable for the electricity and heat manufacturers.

Since the end of XX century the systematic analysis has become the main approach to solve the complex problems. Its application is the basis for a logical and consistent problem detection and validation of the proposed solutions. One of the modern tools of systems analysis for solving the problems of reengineering is a functional simulation using the appropriate computer technology and software [4]. The validity of the application of the functional simulation technology, in particular, is confirmed by ISO 9000 Standards.

Common approaches to coal combustion in the Russian Federation and in the world

According to the first approach, the coal preparation is inefficient, since modern thermal power plant boilers are able to use coal of any quality index. According to the second approach, the feasibility of coal preparation is justified by the increase in TPP electricity production efficiency and depends on many specific factors: the ash and sulfur content, the fuel, transport and coal preparation costs, the environmental protection standards, performance and availability of boilers, etc. [5].

In Germany the combustion of coal with ash-content greater than 2%, is prohibited by the Law. Nowadays, in all industrialized countries 70-90% of all mined stone coal is subject to coal preparation: in the FRG – 95%, in the UK – 75%, in the USA – 55% of coal are subject to coal preparation. South Africa and Australia, preparing 100% of the exported coal, are the world leaders in coal preparation [6].

The increase in ash-content of the low-grade fuel by 1% is:

- the loss of power, the reduction of efficiency factor by 0.1-0.15%;
- the need for additional fuel 0.5 g c.t./kWh;
- the additional suspension in production equal to 60 hours per year for repair of the equipment;
- -the installation of additional equipment for the fuel preparation and ash and slag removal, the increase in capital expenditures;
- the increase in the electricity consumption for own needs by 0.13%;
- the increase in fuel oil consumption per unit of fuel combusted by 0.6-1.5%.

Numerous studies have conclusively proven the benefit of use of the prepared coal in the energy production industry [7-8].

It was found that the elimination of reduced production of electricity at TPP due to a poor coal could cover all the coal preparation costs, even without so much as the benefit from environmental improvements [9].

The construction of modern coal preparation plant located near or directly at the TPP is one of the ways of solving this problem [10].

Functional simulation of the production processes

In 1993, the project was developed and the Standard for the functional simulation (IDEF0) was adopted to meet the challenges of this type [11]. Since the beginning of the XXI century the IDEF0 methodology is used as the primary standardization tool in the domestic economy [12].

The IDEF0 language is standardized in terms of syntax and semantics. The IDEF0 models are three-dimensional, as any two-dimensional IDEF0 diagram can be supplemented by the child functions represented in different layers or levels of the model. The restrictions on the number of child functional layers in the IDEF0 model are absent.

The IDEF0 models are developed and used by the system engineers for simulation and analysis of the complex systems, functional studies of production processes, etc. A number of examples, demonstrating the operational use of IDEF0 are provided both in the domestic and foreign literature [13-15].

In preparing the materials for this article the tools of the free package Ramus Educational [16], similar to AllFusion Process Modeler 7 (formerly BPwin) were used.

The aim of this work is to prove the correctness of the proposed solution of the problem of coal preparation at TPP prior to combustion and to provide an option of the technological chain of operations performed. The solution to these problems has been obtained on the basis of the methodology of functional simulation IDEF0, which is widely used for different types of the systems [17].

The traditional scheme of the construction of the "correct" technological process model was used in this work, according to which the transition from the AS-IS model to TO-BE model was performed [18].

Other methods of systems analysis can be used to solve such problems: the evolutionary algorithms for optimal solutions finding [19], the multiagent simulation [20], the parametric optimization [21], as well as the widespread methods of computational mathematics and mathematical physics [22].

Criteria for the coal combustion efficiency

The integrated approach allows to select a group of criteria for the coal combustion efficiency, represented in the Table 1. The increase in the coal combustion efficiency can be achieved at the expense of its deep preparation for combustion, including the coal preparation and supplementation to activate the process of combustion and disposal of the emissions.

Table 1: Criteria for the coal combustion efficiency

No.	Functional group of criteria	Criteria
1	Performance and technology	– increase in maneuverability of the power units – extension of the adjustment range – reduction of the requirements to the capabilities of the accessories – reduction of the wear and tear – extension of the time between the overhauls
2	Material and technical	– reduction of the reference fuel – reduction of the need for additional highly reactive fuel – reduction of excess air ratio – reduction of the exhaust gases volume
3	Economical	– reduction of the financial costs on exploitation of TPP – reduction of the TPP own needs costs – reduction of the electricity cost
4	Environmental	– reduction of the amount of harmful emissions – reduction of the thermal pollution of the atmosphere – reduction of the soil and groundwater pollution – change in the physical and chemical properties of the solid and liquid waste – improvement of the environmental performance of the electricity product ion process at TPP (gassed space, noise, vibration)
5	Social	– improvement of the working conditions of the TPP staff – creation of the preconditions for the development of the side-line product ions near TPP

Analysis of the methods of mineral raw materials preparation

The preparation processes are relevant to the mining, metallurgical and chemical sectors of industry [23-24]. All methods of preparation are based on the differences in physical and chemical properties of the substances. The first scientific works on mineral raw materials preparation were related to the extraction of the metal ores. Agricola [25] in particular, studied the mechanical methods of preparation such as crushing, washing, and the thermal methods – melting in special furnaces. Lomonosov [26] investigated the separation of the metals from the ores by melting. Modern methods of preparation of the minerals are presented in the Table 2.

Table 2: Methods of mineral raw materials preparation

No.	Used properties of the minerals and additives to be prepared	Groups of methods of preparation	Methods of preparation	Types of methods of preparation
1	Physical	Mechanical	– milling (crushing, grinding) – dispersion (screening)	– pneumatic – hydraulic– in heavy medium
		Gravity	– gravity separation	– pneumatic – hydraulic – in heavy medium
		Thermal	– thermogravitational – pyrometallurgical (roasting, smelting, distillation)	
		Electromagnetic	– electrostatic separation – magnetic separation – electromagnetic separation	– pneumatic – hydraulic
		Radiometric	– radiometric (for uranium ore)	
2	Physical and chemical	Surface phenomena, wettability Thermal	– flotation – thermochemical (melting with the addition of reagents) – Thermo electrolytic (electroly sis)	
3	Chemical	Selective chemical reactions Solubility	– leaching – hydraulic– thermal – halurgical (solubility difference) – extraction (transition in a solution) – method of forming of the poorly soluble compounds (pre cipitation from solution)	
		Other chemical pro cesses	– hydrolysis – electrolysis – ion exchange method	
4	Biochemical		– bacterial – microbiological	

Analysis of the technological processes

In the late 90-ies of the XX century, most of the coal preparation plants in Russia were closed. This led to a sharp increase in the ash-content of the coal combusted. This situation provides extremely negative impact not only on the performance of the thermal power plants, but also on the environment nearby the power plant. In terms of the macro-economy, it is advisable to perform the coal preparation close by the places of its production with the subsequent sale to the consumers. In the context of the slow rebirth of the centralized coal preparation and sale of the non-prepared coal of variable ash-content to the consumers, it seems appropriate to the authors to include the coal preparation stage in the process of preparation of the air-coal dust mixture (Fig. 1). The proposed innovations are reflected in the diagram TO-BE of the process of coal preparation for combustion (Fig. 2). This diagram also reflects the changes in the process of supply to the boiler, associated with the deep coal preparation for combustion, including, except the preparation, the introduction of additives (activators, catalysts), activating the combustion process.

The context diagram of the TO-BE process of the coal preparation for combustion (Fig. 3), in addition to the described input, output streams and mechanisms for the implementation of the technological operations, shows a list of control and regulatory documents, setting the conditions for the implementation of the technological and logical processes, described in the model.

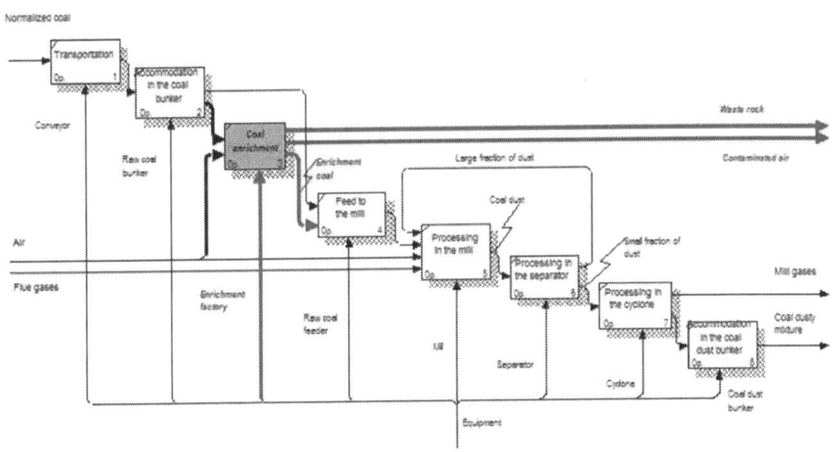

Fig. 1 – Diagram of the process of preparation of the air-coal dust mixture.

Figure 1: Diagram of the process of preparation of the air-coal dust mixture

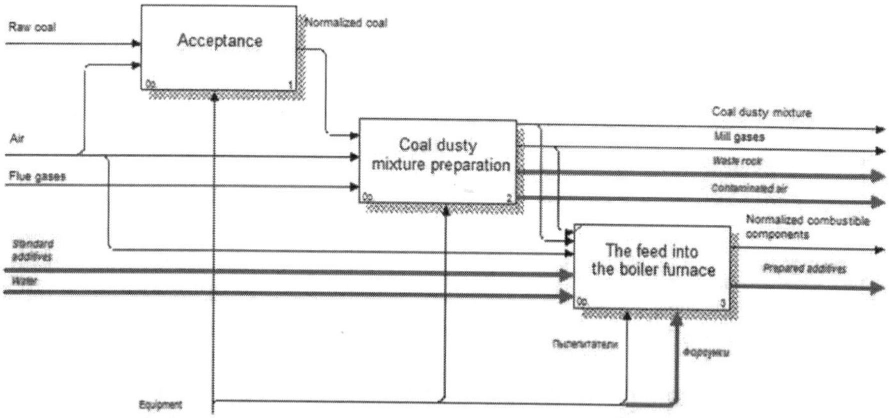

Fig. 2 – Diagram of the TO-BE process of the coal preparation for combustion.

Figure 2: Diagram of the TO-BE process of the coal preparation for combustion

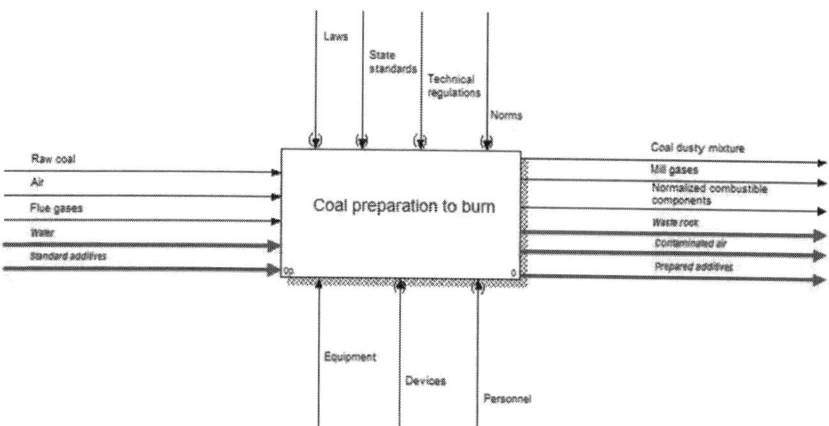

Fig. 3 – Context diagram of the TO-BE process of the coal preparation for combustion.

Figure 3: Context diagram of the TO-BE process of the coal preparation for combustion

Coal preparation equipment, available on the special equipment market

The analysis of the literature and the Internet sources allowed to identify the main manufacturers of the coal preparation equipment, available on the market of the Russian Federation (Table 3).

Table 3. Manufacturers of the coal preparation equipment

No.	Country of origin	Countries and regions-consumers	Trademarks of the equipment	Web-site of the supplier
1	China	China Europe, USA, Russia, Ukraine, Brazil, Turkey, South Africa, Ka zakhstan, Iran, North Korea, the Philippines, Viet Nam, Indonesia, Mongolia	Batak, Baum, FGX	Equipment from China, www.china-bridge.ru
2	Ukraine (Lug ansk)	Ukraine, Russia	SκV, Sκ, SÊVP, GP etc.	Alliance Oil, al yansoil.co m
3	Germany	Europe Russia	al lair®Allmineral, CETCO	Vibrotechtsentr, www.vtcenter.ru
4	China	Southeast Asia, Eastern Europe, South America, Middle East, Africa	Liming	Liming LLC, Shanghai Mining Machinery Company, fakmil.com
5	South Africa (including under license of Telsmith, American Pulverizer (USA), Mogensen (United Kingdom), IFE (Austria))	South America, Asia, Australia, Russia, CIS countries	Osborn	Ore and rock proc essing, www. gorniedrobilki.com
6	Russia (Novosibirsk)	Russia, CIS countries	Sepair	Gormasheksport, gmexp.ru

Selection of the coal preparation plant according to the method of T. Saaty

The evaluation of the different variants of the technological equipment was performed based on the hierarchies analysis method of T. Saaty [27-28].

Proceeding from the foregoing list of equipment and its characteristics (Table 3) the list of alternative equipment brands including five positions was completed as follows: FGX (China), SKV (Ukraine), allair®Allmineral (Germany), Osborn (South Africa) and Sepair (Russia).

The criteria for selection of the equipment were determined on the basis of the criteria of the coal combustion efficiency (Table 1): K1 shows the efficiency (the assessment by this criterion involves the proximity of the equipment index to the desired value for the power unit of 300 MW); K2 shows the dimensions (the criterion is focused on the possibility to mount the coal preparation unit near the fuel and transport workshop of TPP); K3 shows the energy consumption (this criterion takes into account the consumption of the electricity by the main and auxiliary machinery of the unit); K4 shows the water consumption (taking into account the need to use water resources); K5 shows the adaptation (this criterion indicates the extent to which the technology used in the unit, complies with the domestic environmental conditions and standards).

The ratings (priorities) for the criteria are set in the range of 1 to 9 where 1 means the equal importance of the criteria, 9 means a very severe superiority of one criterion over the other. The calculation results are presented in the Table 4. The best option of the processing plant is the option number 5 (Sepair, Russia), having a maximum value of a global priority. The option number 2 (SKV, Ukraine) should be noted, which is the second (close) preferred alternative.

Table 4. The results of the evaluation of the alternatives

No.	Alternatives	Criteria					Global
		K1	K2	K3	K4	K5	priorities
The numerical value of the priority vector							
		0.029	0.044	0.128	0.273	0.526	
1	FGX	0.108	0.053	0.126	0.222	0.176	0.175
2	SKV	0.065	0.456	0.118	0.111	0.313	0.232
3	allair®Allmineral	0.042	0.089	0.100	0.222	0.099	0.130
4	Osborn	0.108	0.152	0.095	0.222	0.099	0.135
5	Sepair	0.677	0.250	0.561	0.222	0.313	0.328

The "Sepair" product line includes the industrial unit with the capacity of 120 t/h of raw coal, while the demand of the coal energy unit of 300 MW is 100 t/h. The coal with the original ash-content of 22-23%, class 14-25 and 25-40 mm is divided by the "SEPAIR" unit in the low-ash coal (ash-content of 8-12%), traded coal (16-18%), middling (35%) enclosing the high-ash solids (tailing, ash-content of 75% or more). The schematic diagram of the "Sepair" unit is shown in the Fig. 4.

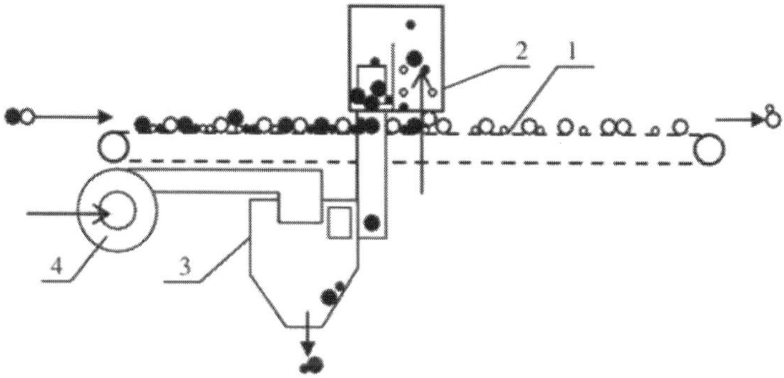

Fig. 4 – Schematic diagram of the "SEPAIR" pneumatic separator:
1 – belt wire-net conveyor; 2 – air nozzle;
3 – product bin; 4 – blower.

Figure 4. Schematic diagram of the "SEPAIR" pneumatic separator

Technical and economic effect of the modernization of the facility

The technical and economic analysis of the proposed solutions showed, that for 300 MW power unit at the cost of the coal preparation unit equal to 300 million rubles, the annual fuel consumption equal to 878,400 tons and the fuel price equal to 2000 rub./t:

- the capital investments, including the transportation costs, building and construction works, will amount to 456 million rubles;
- the annual operating costs, including the depreciation deductions and the cost of the annual maintenance, will amount to 41.04 million rubles;
- the annual savings from the reduction of the fuel consumption will be 10,600 tonnes/year, or 21.2 million rubles;
- the reduction of fees for sulfur oxides [29] emissions will amount to 37.4 million rubles, while the volume of emissions after the event will be reduced by 50%;

- the total annual savings net annual operating costs, related to the coal preparation unit, will be equal to 17.56 million rubles;
- the net income, subject to tax on profits equal to 20%, will amount to more than 14 million rubles;
- the simple payback period, taking into account the net income, depreciation and amortization deductions, will be approximately 12 years from the commissioning of the coal preparation unit "SEPAIR", but in the case of a modular (simplified) performance of the coal preparation line, by reducing the time and the installation costs, the cost of the coal preparation will be reduced to 25-30 rubles per tonne of raw coal [30], which will allow to reduce the payback period to 2 years.

Conclusions and recommendations

To solve the problem of high-ash coal combustion efficiency improvement at the thermal power plants it is necessary to develop a new approach to the evaluation of the coal washability as well as the concept of rational use of the steam coal, defining the fundamental improvement in the efficiency of the coal power by the substitution of the mine coal by the prepared coal in the fuel balance of the thermal power plants and the high competitiveness of the prepared coal fuel, commensurate with the other organic energy carriers. These measures will allow to create the technology of environmentally friendly coal-burning fuel and energy complexes. To this end, the authors plan the further development of the simulation models, based on the agent approach to the thermal power plants for the rapid assessment of the effectiveness of the solutions, similar to the solutions, proposed in this work, for the specific power plants, and, possibly, the regions.

The work was financially supported by the Ministry of Education and Science of the Russian Federation in the framework of the implementation of the FTP "Research and development on the priority development fields of the scientific and technological complex of Russia for 2014-2020" (Agreement No 14.574.21.0018 on "Development of the method for the intensification of the low-reactive coal gasification process in an upward jet-vortex of the oxidant." Unique identifier PNI (of the Project) RFMEFI57414X0018).

REFERENCES

1. *Energy Strategy of Russia until 2030 (App. to the public and business magazine "Energy Policy")* (p. 184). (2010). Moscow: Institute of Energy Strategy.

2. *GOST R 51586-2000. Brown Coals, Hard Coals and Anthracites of Kuznetsk and Gorlovky Basins for Power Supply Purposes. Specifications.*(2000). Retrieved June 5, 2015, from http://docs.cntd.ru/document/1200026841.

3. *GOST 17321-71. Coal. Preparation. Terms and Definitions.* (1971). Retrieved June 5, 2015, from http://standartgost.ru/.

4. Cheremnyh, S.V., Semenov, I.O., and Ruchkin, V.S. (2002). *Simulation and Systems Analysis: IDEF-Technology: Practical Course* (p. 192). Moscow: Finance and Statistics.

5. Holt, E.C. Jr. (1982, May). Effect of Coal Quality on Maintenance Costs of Utility Plants. *American Mining Congress Journal.*

6. *The Official Website of Mining Industry Journal.* (n.d.). Retrieved June 6, 2015, from http://www.mining-media.ru/.

7. *SibAcademInnovatsiya website.* (n.d.). Retrieved June 6, 2015, from http://www.sibai.ru/.

8. Burdukov, A.P., Popov, V.I., Faleev, V.A., and Yusupov, T.S. (2010). The Use of the Mechanically Activated Coal of Micropulverisation in the Energy Sector. *Polzunovskiy Vestnik, 1,* 93-98.

9. Gorshkov, A.S. (1983). The National Economic Damage Caused by the Deterioration of the Steam Coal Quality and Coal Preparation Efficiency. *Electricheskie stantsii, 9,* 2-4.

10. Fedorova, N.V., and Krivobok, E.A. (2015). Preparation of High-Ash Coals to Combustion at the Thermal Power Plants. *Ecologiya Promishlennogo Proizvodstva, 1*(89), 41-46.

11. Air Force Wright Aeronautical Laboratories. (1981). *ICAM architecture. Part II, Volume IV. Function modeling manual (IDEF0)*. Ohio: Air Force Systems Command, Wright-Patterson Air Force Base.

12. *Methodology for Functional Simulation*. (2000). Moscow: Rossiyskiy Gosudarstnenniy Standart.

13. Komissarova, M.A., and Leonova, S.A. (2014). Improvement in the Effectiveness of the Coal Mining Companies Based on the Process Simulation. *Gorniy Informatsionno-analiticheskiy Bulleten, 8*, 227-284.

14. Cochran, L., and Wheaton, K. (2002). *A Conceptual Operational Model for Command and Control of International Missions in the Canadian Forces.* Ottawa: FirstMark Technologies Ltd.

15. Selivanov, S.G., and Gavrilova, O.A. (2014). Development of the Innovative Methods of Production and Optimization Methods for the Aircraft Engine Technology. *Vestnik UGATU, 4*(64), 61-71.

16. *Ramus – a Cross-Platform System for Simulation and Analysis of Business Processes.* (n.d.). Retrieved July 5, 2015, from http://ramussoftware.com/.

17. National Institute of Standards and Technology. (1993). *Announcing the Standard for Integration Definition for Function Modeling (IDEF0).*

18. Waissi, G.R., Demir, M., Humble, J.E., and Lev, B. (2015). Automation of Strategy Using IDEF0 – A Proof of Concept. *Operations Research Perspectives, 2*, 106-113.

19. Kuznetsova, A.V., Mokhov, V.A., and Turovskaya, E.V. (2014). Simulation of Optimal Placement of Goods at the Self-Service Stock Based on the Evolutionary Search Algorithms. *Inzhenerniy Vestnik Dona, 1*. Retrieved August 25, 2015, from http://www.ivdon.ru/magazine/archive/n1y2014/2276.

20. Mokhov, V.A., Gheorghitsa, I.V., and Goncharov, S.A. (2013). Multi-Agent Simulation of the Network Attack Such as DDoS. *Inzhenerniy Vestnik Dona, 3*. Retrieved August 25, 2015, from http://ivdon.ru/magazine/archive/n3y2013/1852.

21. Mokhov, V.A, and Borodulina, E.N. (2014). On the Issue of Parametric Optimization of the Swarm Algorithms. *Izvestiya SFU. Technicheskie nauki, 4*, 230-234.

22. Efimov, N.N., and Tyrnikova, U.V. (2007). Method of Calculation of the Temperature Field of the Fin Tubes of the Evaporator Heating Surfaces of the Boilers. In *Improving the Efficiency of the Electricity Production: Materials of the VI International Scientific and Engineering Conference* (pp. 240-243). Novocherkassk: Onyx+.

23. *Promyshlennaya ekologiya website*. (n.d.). Retrieved June 5, 2015, from http://prom-ecologi.ru/.

24. *Vse o Gornom Dele website*. (n.d.). Retrieved June 5, 2015, from http://computerchoppers.ru/.

25. Agricola, G.O. (1986). *On Mining and Metallurgy* (p. 292). Moscow: Nedra.

26. Lomonosov, M.V. (1763). *The Fundamentals of Metallurgy or Mining* (p. 436). St. Petersburg: Publ. House of the Imperial Academy of Sciences.

27. Saaty, T.L. (2008). *The Decision-Making at the Dependencies and Feedback: Analytical Networks* (p. 360). Moscow: Pubishing house LKI.

28. Saaty, T.L. (1987). Concepts, Theory and Techniques: Rank Generation, Preservation and Reversal in the Analytic Hierarchy Process.*Decision Sciences, 18*, 157-177.

29. Novikova, Zh.I. (2008). How to Calculate the Cost of Pollution. *Rossiyskiy Nalogoviy Kurier, 6*. Retrieved June 6, 2015, from http://e.rnk.ru/article.aspx?aid=231984.

30. *Dry Preparation – SEPAIR Air Separation Unit.* (n.d.). Retrieved November 6, 2015, from http://gmexp.ru/equipment/type/grav/sepair.html.

5

CHAPTER

Feasibility Studies on Cleaning of High Sulfur Coals by Using Ionic Liquids

Binoy K. Saikia[1] • Kakoli Khound[1] • Om P. Sahu[2] • Bimala P. Baruah[1]

[1]*Coal Chemistry Division, CSIR-North East Institute of Science & Technology, Jorhat, Assam 785006, India*

[2] *Analytical Chemistry Division, CSIR-North East Institute of Science & Technology, Jorhat, Assam 785006, India*

ABSTRACT

Coal has been used as an energy resource around the world, primarily for the generation of electricity. The cleaning of coal by removing its unwanted sulfur and mineral matter components is utmost essential before their gainful utilizations. The ionic liquids (ILs) are considered as non-toxic solvents for using in different industrial processes. The effect of two ILs namely, 1-*n*-butyl, 3-methylimidazolium tetrafluoro borate (IL1) and 1-*n*-butyl, 3-methylimidazolium chloride (IL2) in oxidative de-sulfurization and de-ashing of two industrially important high sulfur coal samples from Meghalaya (India) is discussed in this paper. The maximum removal of total sulfur, pyritic sulfur, sulfate sulfur and organic sulfur are observed to be 37.36 %, 62.50 %, 83.33 % and 31.63 % respectively during this oxidative process. The quantitative diffuse reflectance Fourier transform-infrared spectroscopy analysis supports the formation of sulfoxides (S=O) and sulfones ($-SO_2$) and their subsequent removal during the oxidation of the coals in presence of ILs. The X-ray fluorescence combined with near edge X-ray absorption fine structure and scanning electron microscopic studies reveal the removal of mineral matters (ash yields) from the coal samples. The thermogravimetric analysis of the raw and clean coals indicates their high combustion efficiencies and suitability for using in thermal plants. The method is partially green and the ILs could be recovered and reused in the process.

Keywords: High sulfur coals De-sulfurization of coal Indian coals Ionic liquids De-ashing

1 INTRODUCTION

Coal has been used as the major source for electricity generation in many parts of the world including India. India is the third largest coal producing country in the world and 'Coal India Ltd.' is one of the world's largest coal mining companies (Stracher and Taylor 2004). There are over 861 billion tonnes of proven coal reserves worldwide, which is enough to last for around 100 years at current rates of production. However, cleaner approaches are vital for gainful and sustainable utilizations of coals. Coal naturally contains sulfur and mineral matters (ash yield). Sulfur in coal is generally grouped into inorganic and organic forms. The high sulfur coals create environmental problems to the community and thus, minimization of sulfur and mineral matters from coals is very much essential before their various types of utilizations. The acid mine drainage (AMD) is also caused by the oxidation of pyritic sulfur in coals (Baruah et al. 2006). The performance of a coal-based power plant is also influenced by many coal properties. The mineral matters and sulfur contents of any coal have most impact on the boiler performance. The presence of mineral matter (ash contents) can cause ash deposition and erosion damage to the boiler tube. The acid dew point of flue gas is also determined by the sulfur content of the coal. Higher coal sulfur level may promote air heater corrosion (Juniper and Pohl 1997). In steel making industry, the sulfur content of coke, sediments on the iron crystal surface make steel more brittle and decrease its plasticity property (Ehsani 2006). In recent years, chemical desulfurization of coals has attracted much attention due to the gradual reduction of the statutory sulfur contents. A large number of chemicals have been used to remove the sulfur and mineral matters from coals (Yaman et al. 1954; Mazumdar et al. 1989; Sain et al. 1991; Demirbas and Balat 2004; Komintarachat and Trakarnpruk 2006; Baruah and Khare 2007; Ahmed et al. 2008; Zhang et al. 2009; Campos-Martin et al. 2010; Silva et al.2011). The most common industrial process of catalytic hydro-desulfurization (HDS) needs hydrogen with additional high investments and operation cost (Schmitz et al. 2004). On the other hand, the chemicals used during the desulfurization processes also have been creating environmental hazards to the communities as well as surroundings. As a consequence, chemists are exposed to a large number of lethal chemicals during their working. Thus, the chemical desulfurization process must be compatible to the environmental restriction as well as to the chemists. So, suitable and environment friendly chemical desulfurization method is urgently needed for the coal processing plants. In the last few decades, there emerged a new class of green solvents known as "Ionic Liquid" (IL) (Rogers and Seddon 2003). They are composed of cation and anion, non-volatile, excellent solubility, great thermal/chemical stability and non-flammability and environmental friendliness. The ILs have been used potentially in different sectors including nuclear applications to the petroleum industry. The ILs have also been reported for direct extraction of S-compounds (extractive desulfurization, EDS) from liquid fuels (Bosmann et al. 2001; Esser et al. 2004; Zhao et al. 2005; He et al. 2008; Holbrey et al. 2008; Kendra-Krolik et al. 2011).

The coal resources of India are available in older Gondowana formation of Peninsular India and younger to tertiary formation of North-eastern region (NER). The NER of India has a large reservoir of high sulfur (2 %–8 %) coals with sub-bituminous in rank (Saikia et al. 2014). The majority of sulfur functionalities of these coals were found to be in the organic forms (75 %–90 %), which are retardant to remove chemically (Baruah and Khare 2007). There are five types of organic sulfur functionalities found in these NER coals including aliphatic or aromatic thiols, aliphatic or mixed sulfides, aliphatic or aromatic disulfides, heterocyclic compounds of thiophenic and thioketonic types, etc., (Barooah and Baruah 1996); moreover, the NER coals are characterized by low ash and high caking index, suitable for use in cement plant and metallurgy as coking coal. But, these coals cannot be gainfully utilized because of their high sulfur contents as mentioned above.

We present here a partially green chemical beneficiation method for de-sulfurization and de-ashing of high sulfur low grade NER coals from Meghalaya (North-east India) by using ILs as non-toxic solvents in a mild reaction conditions with no need of hydrogen, pressure and reactor. The prime objective of this study is also to see the outcome of the oxidative desulfurization (ODS) process with those high sulfur coals in presence of the ILs. The ODS process includes oxidation of sulfur compounds followed by their extraction in solvents. The use of green solvents like ILs in this present coal cleaning process for high sulfur coals makes the process partially green and efficient. The process may also be considered in clean coal technology.

2 MATERIALS AND METHODS

The chemicals (Sigma-Aldrich and Merck) used were: hydrogen peroxide (30 %), formic acid (85 %), 1-*n*-butyl,3-methylimidazolium tetrafluoro borate (IL1) and 1-*n*-butyl,3-methylimidazolium chloride (IL2). The representative high sulfur coal samples (SCF and B) were collected from two different collieries of Bapung coalfield (N 27°13′–27°23′; E 95°35′–96°00′), Meghalaya (India) and ground to −0.211 mm sizes for other analyses by suing standard method (ASTM 2010). The areas under investigation is covered by the Barail group of rocks consist coarse sandstone, shale, minor coal lenses and carbonaceous shale. The coal seams are associated with the Lakadong Sandstone of Shella Formation of Jaintia group (Eocene) includes three persistent coal seams ranging from 1–3 m in thickness along with inter-seam sandstone and shale (Ahmed and Rahim 1996).

The proximate analysis and sulfur analysis were carried out in Proximate Analyser (TGA701, Leco) and Sulfur Analyser (LECO, 144DR) respectively by using standard method (ASTM 1991). The forms of sulfur were determined by following the standard method (ASTM 2005). The percent of organic sulfur was calculated by difference. The gross calorific values (GCV) were also determined in Bomb calorimeter (Model: Leco, USA) by using standard methods (ASTM 2013).

2.1 Quantitative diffuse reflectance FT-IR spectroscopic analysis of coals

Diffuse reflectance Fourier transform infrared (FT-IR) spectra of the coal samples were recorded using FT-IR instrument (System-2000, Perkin–Elmer Inc.) in the range of 500–1500 cm^{-1}. The samples (about 6–7 mg) were accurately weighed on single pan electronic balance (Model: SI-234, Denver Instrument) and thoroughly mixed with spectroscopic grade KBr powder (about 200 mg) in an agate mortar. Pure ground KBr was used to obtain the reflectance spectrum by taking 64 numbers of scan having resolution 4.0 cm^{-1}. The FTIR spectra were processed using the software Spectrum V-3.02 (Version 3.02.01, Perkin–Elmer) in absorbance mode. At first the base line was corrected and was brought to the 0.0 reading. The spectra were normalized to 1 mg of the samples by dividing it with the amount of the sample taken for recording the drift spectra to compare the absorbance at 563 cm^{-1}.

2.2 Thermal (TG–DTG), XRF and SEM analysis of coals

Thermogravimetry (TG) and their first derivative thermogravimetric (DTG) spectra were recorded in the thermal analyzer (Model: TGA701, LECO) with heating rate of 10 °C/min in presence of oxygen environment. The combustion performance index (S) and energy of activation (Ea) were calculated by using the methods as reported elsewhere (Sonibare et al. 2005; Qing et al. 2011).

The high energy X-ray fluorescence (XRF) measurements of coal samples before and after the treatment were obtained from the Canadian Light Source [Saskatchewanin the Soft X-ray micro-characterization Beamline (SXRMB)]. The Si(111) crystals were used for XRF measurements with photon energy set at 8350 eV. The powder sample was mounted using double-sided, conducting carbon tape, and loaded into a vacuum chamber (base pressure of 1×10^{-8} torr). The scanning electron microscopic (SEM) images were recorded in SEM (Model: JSM-6360 JEOL).

2.3 Removal of sulfur components and mineral matters (ash yield)

Thirty grams of coal sample were refluxed with a mixture of 13.2 mL of 20 % formic acid and 75 mL of 20 % H_2O_2 for 1 h. Then, the oxidized coal samples were filtered and washed with water until the pH of the filtrate became neutral. In another two separate experiments, 30 g coal samples (0.211 mm in size) were oxidized with the mixtures of 13.2 mL of 20 % formic acid and 75 mL of 20 % H_2O_2 in the same reflux conditions for 1 h in presence of the IL1 and IL2 (0.5 g for each) respectively. The coal-IL mixtures were centrifuged and the recovered coal residues were washed with hot distilled water until neutral and oven dried for 1 h. The filtrates obtained were kept for the recovery and further use of the ILs. The oven dried oxidized coal samples were used for subsequent analysis.

The filtrates obtained in the above step were centrifuged with absolute alcohol to separate out any impurities from the mixtures. Absolute alcohol was separated by traditional evaporation and then, the filtrates were mixed with 1 % $BaCl_2$ solution followed by CCl_4 (about 50 mL) to precipitate out dissolved sulfur functionalities. White colored substances were formed upon slight heating at about 60 °C due to the precipitation of inorganic and organic sulfur compounds, which were removed from coals during oxidation. The white precipitates were then filtered through Whatmann-42 filter paper. The final filtrate containing ILs became colorless. After boiling off the CCl_4, the filtrates could be used for another desulfurization experiments with slight decrease in activity.

3 RESULTS AND DISCUSSION

3.1 Physico-chemical characteristics of coals

The physico-chemical properties of the raw and oxidized coal samples are shown in Table 1. It demonstrates that the SCF and B coals are high sulfur and high ash in nature. Moreover, the majority of sulfur is organically bound in the coal samples (up to 79.0 %). The geochemical studies of sulfur in coals provide valuable information regarding the sources of sulfur and processes of sulfur incorporation into coals which helps in the study of the depositional environments and formation history of coals (Chou2012). The major part of inorganic sulphur in these coals consists of pyrite (FeS_2), which is usually in particulate, to agglomerated forms (Baruah et al. 2006). The high sulphur content in the Northeast Indian coals resulted from the increased availability of sulphate ions in sea water coupled with the activity of anaerobic bacteria. The presence of high organic sulfur in northeast Indian coals suggests the marine influence as sulfur compounds are quickly formed by oxidation in marine environment. The oceanic sulfate sources have high sulfate reduction rates and during the coal formation, these sulfates are incorporated to the matrix of the coal molecule giving high organic sulfur groups. The geochemistry of the high sulphur Northeast Indian coals are reported elsewhere (Nayak et al. 2009; Saikia et al. 2014).

Table 1. Physico-chemical analysis of the raw and beneficiated coal samples (as received, %wt)

Coals	M	Ash	VM	FC	TS	PS	SS	OS	%PS removal	%SS removal	%OS removal	%TS removal	%Ash removal	GCV (kcal/kg)
SCF	4.07	25.9	31.8	38.3	4.47	0.80	0.12	3.6	–	–	–	–	–	6565.60
SCF Oxd	2.58	22.3	32	43.12	3.37	0.60	0.06	2.6	25.0	50.00	27.77	24.60	13.89	7155.00
SCF +IL1	2.10	23.5	33.3	41.10	3.25	0.55	0.04	2.6	31.25	66.66	26.11	37.36	9.26	7411.79
SCF + IL2	2.25	25.6	32.4	39.7	2.80	0.30	0.02	2.48	62.50	83.33	31.11	27.29	1.15	7780.60
B	4.30	17.8	33.7	44.0	3.47	1.08	0.67	1.77	–	–	–	–	–	6687.56
B Oxd	2.34	15.9	36.9	44.9	2.84	0.75	0.45	1.65	30.55	32.83	7.34	18.15	10.67	6812.08
B + IL1	4.72	15.5	33.7	46.08	2.71	0.98	0.52	1.21	9.25	22.38	31.63	21.90	12.92	7140.12
B + IL2	2.31	16.0	37.4	44.3	2.36	0.65	0.60	1.73	39.81	10.44	2.25	14.12	10.11	7213.40

M moisture, *VM* volatile matter, *FC* fixed carbon, *TS* total sulfur, *PS* pyritic sulfur, *SS* sulfate sulfur, *OS* organic sulfur

3.2 Removal of sulfur and mineral matter (ash yield)

As observed, the oxidation these coal samples (B and SCF) in ILs (IL1 and IL2), the pyritic, organic and sulfate sulfur contents have been significantly removed. On oxidation, the maximum removal of total sulfur and mineral matters (ash yield) of the coal samples are found to be 37.36 % and 13.89 % respectively. However, the pyritic, sulfate and organic sulfur are significantly removed by 62.50 %, 83.33 % and 31.63 % respectively during the process. Moreover, the fuel values (i.e., GCV) are also found to be increased after the beneficiation process (Table 1). The Fig. 1 shows the effectiveness of the process in removal of the mineral matters (ash yield) and sulfur components from the two coal samples. It is observed that the present process of ODS is more effective for the SCF coal sample than that of B coal although they are from the same coalfield. Thus, the removal of sulfur including pyritic, sulfate and organic sulfur varies from one colliery to other in the same Bapung coalfield. It may be due to the difference in organic sulfur functional groups present in the coal structure. However, it needs geo-chemical study to understand these variations in the same coalfield. The oxidation in presence of ILs produces sulfone with other components of the product being controlled by factors such as reaction temperature and pH (Davalos et al. 1994). Some researcher found that S-removal in ODS does not linearly increases with H_2O_2 dosage in IL (Zhao et al. 2010). The results were reported that the sulfoxide functional groups decreases with coal rank, while the contents of thiophene increases (McCoy and Weiss1954; Calkins 1994; Yperman et al. 1995; Maes et al. 1997). Thus, the stable thiophenes are formed from less stable organic sulfur compounds (thiols and sulfides) with increasing rank. On the other hand, coal of different ranks/types has a different distribution of organic-sulfur functionalities, which may affect the extent of desulfurization.

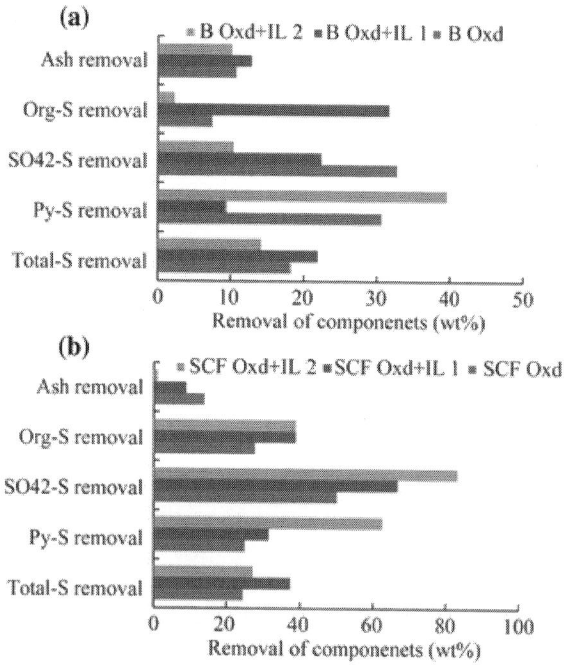

Figure 1. Extent of removal of components during oxidation of coals with ILs

In the oxidative process, the acidic ILs decomposes H_2O_2 to form OH, which is strong oxidant for organic S-compounds present in the coals. The organic sulfur compounds were selectively oxidized into sulfone and sulfoxide in the IL phase. Generally, the divalent S-atom of the organic sulfur compounds undergoes the electrophilic addition of oxygen atom from the H_2O_2 to form sulfone during oxidation. The present Bronsted acidic ILs is water soluble and promotes catalytic reactions; hence in presence of hydrogen peroxide they will have more oxidizing effect. However, the strength and extent of coal oxidation may vary with the rank/type of coal and nature of ILs.

3.3 X-ray diffraction analysis of coals

It is to be mentioned that the low removal of mineral matters (de-ashing) were observed during this process (Table 1). The similarities of the XRD profiles of the raw and treated coals also support the low removal of mineral matters (Figs. 2, 3). The associated minerals present in the raw and oxidized coals are also shown in Figs. 2 and 3. Meghalaya coals of the NER (India) are of Gondowana type, where the majority of mineral matters are present in the inertinite parts. It has higher density and hard to remove and hence cannot be separated by mild and conventional ash removal methods (Falcon and Ham 1988). However, the pyrite of these coals is in the vitrinite part, which is removable easily after oxidation reactions.

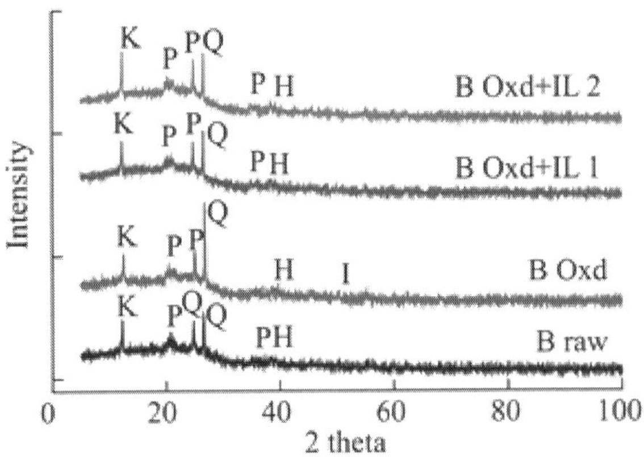

Figure 2. X-ray diffractogram of raw and beneficiated B coal sample (*K* kaolinite, *P* pyrite, *Q* quartz, *H* hematite, *I* illite)

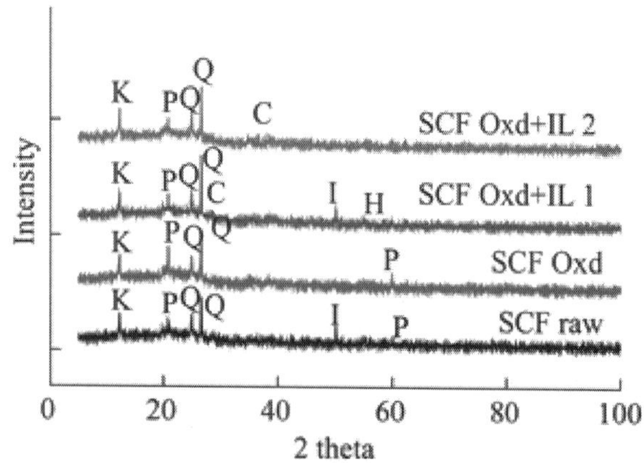

Figure 3. X-ray diffractogram of raw and IL beneficiated SCF coal sample (*K* kaolinite, *P* pyrite, *Q* quartz, *C* calcite,*H* hematite, *I* illite)

3.4 Quantitative FT-IR spectroscopic analysis of sulfur transformation

The FT-IR spectra of the raw and oxidized coals are shown in Figs. 4 and 5 and it is observed that the amounts of sulfur functionalities are different in both the coals. The absorption bands in the region of 550 cm^{-1} are assigned to the stretching vibration of S–S bonds of sulfides and C–S bonds in primary and secondary thiols (Silverstein et al. 1987). The bands in the region of 1415–1380 and 1200–1185 $cm^{-1}$are observed due to the presence of organic sulfates. The stretching frequencies in the range of 1070–1030 cm^{-1} are due to the presence of sulfoxide in the coal samples. The bands in the range of 1050–1020 cm^{-1} are due to the formation of S=O in the oxidized coal samples (Silverstein et al. 1987; De Filippis and Scarsella 2003). However, these regions are also seen in the raw coal sample due to the presence of sulfate ($-SO_4$). The FT-IR spectra of the raw and oxidized coals were normalized and it is interestingly observed that the absorbance of the sulfoxide, sulfones and sulfate groups were decreased appreciably during the total beneficiation process. Thus, the FT-IR analysis reveals that the sulfur has been converted to sulfoxide (S=O) and sulfone ($-SO_2$) on oxidation and subsequently removed by ILs. In this study, the extent of desulfurization differs due to the difference in sulfur contents in the coal samples.

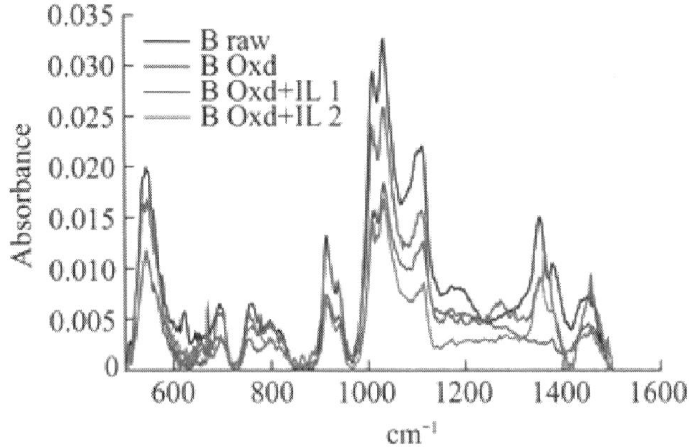

Figure 4. FT-IR spectra of the raw and beneficiated coal samples (B) in IL1 and IL2.

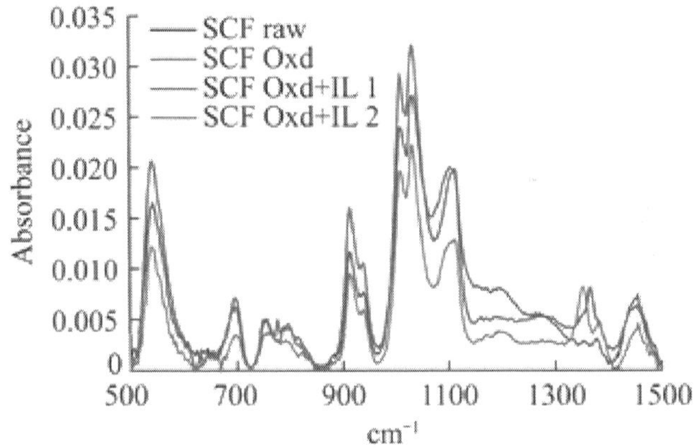

Figure 5. FT-IR spectra of the raw and beneficiated coal samples (SCF) in IL1 and IL2

3.5 X-ray fluorescence analysis on mineral matter (ash yield) removal

The XRF combined with the NEXAFS analysis of the raw and beneficiated (oxidized) coal samples are depicted in Fig. 6. The excitation energy has been chosen just above potassium. All XRF spectra were normalized to the same intensity at the excitation

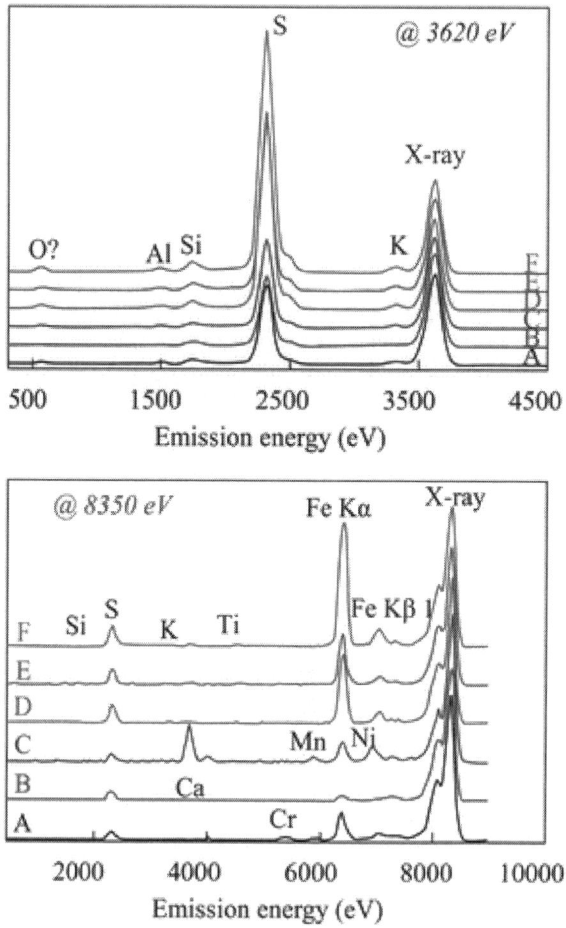

Figure 6. XRF Spectra of the raw and beneficiated coal samples at two different eV (*A* SCF raw, *B* SCF Oxd, *C* B raw, *D* B Oxd, *E* B Oxd + IL1, *F* B Oxd + IL2)

energy (scattered X-rays), so it is easy to say that the current beneficiation process converts the sulfur functionalities to sulfones and sulfoxide (Fig. 6f). The XRF could be useful in showing: (a) the change in sulfur intensity due to oxidation; (b) change in inorganic composition, particularly in for oxidation followed by IL extractions (Fig. 6c–f); (c) the B sample has more sulfur than SCF sample, which is in agreement with the chemical analysis. It has also a lot more Ca,

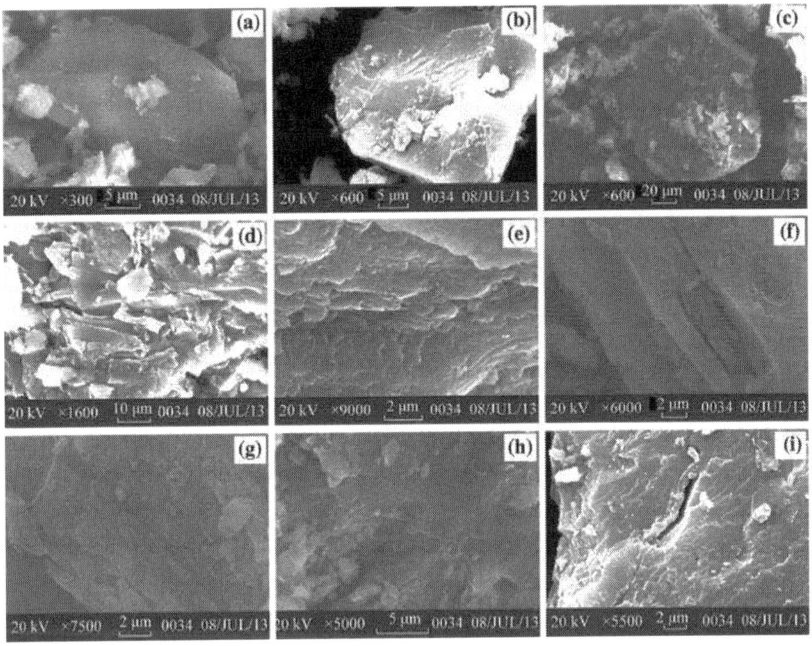

Figure 7. SEM (SE) morphology of the raw and beneficiated coal samples (**a–c** raw coals; **d–i** IL-treated coals)

Mn and Ni along with pyrite and sulfate as seen in the XAFS; (d) for SCF coal: most of the inorganic species have been removed/reduced after the treatment indicating the maximum removal of ash (13.09 %); (e) for both the coals, it seems that sulfur peak has increased after the treatment, which could be due to the conversion of sulfur forms to inorganic sulfur, such as Fe_2S and $FeSO_4$, as seen in the increase of the Fe intensity. In addition, the ILs treatment of B coal removed most of other inorganic species such as Ca, Mn and Ni.

3.6 SEM analysis of raw and beneficiated coal samples

The effect of H_2O_2, HCOOH and ILs on the microstructure of the coals is displayed in Fig. 7, which compares the changes in morphologies of the raw and beneficiated coal samples. The SEM micrographs of the beneficiated coal samples (Fig. 7d–h) reveal the occurrence of leaching of the coals during oxidation. The appearance of small hollow pits on their surface might be due to the disappearance of pyritic, sulfate and mineral matters from the coal samples on oxidation. A few small and large particles (in Fig. 7d–g) are found to be dispersed in the beneficiated coals. Thus, the changes in the SEM morphologies support the present beneficiation process through removal of sulfur components and mineral matters from the coal samples (Ishak and Ismail 2004).

3.7 Thermal properties of the coals

The TG–DTG profiles of the raw and beneficiated coals mainly indicate four temperature ranges (Fig. 8). The DTG curve express the starting and closing temperatures of the decomposition and the sharp peak gives the point of decomposition of the coal samples. The residual moisture releases at temperatures range of 25–100 °C. In this stage, the rate of weight loss increases as observed in the TG–DTG curves (Fig. 8). It is observed that the volatile matters start to be released along with the reduction of H-bonding around 223 °C. The weight loss is also increased at this stage as revealed from the TG curve of the both coal samples. However, both the beneficiated coal samples (oxidized) show significant release in volatile matters at this stage temperature. In the third stage (at higher temperature around 400 °C), the fixed carbon in coals begins to oxidize (releasing CO and H_2) and the volatile matters are completely released and the weight loss has been sharply increased. The functional groups, like R–O–R, R–OH present in beneficiated coals facilitates greater decomposition at lower temperature that are released as CO and CO_2. However, the C–H bond present in the raw coal samples are difficult to break than the C–O bond present in the oxidized coal samples (Saikia et al. 2009). Thus, the burning temperatures of the beneficiated coals are reduced indicating their better combustion efficiency. At a temperature range of 500–600 °C, the decomposition of ash occurs e.g., kaolinite → metakaoline at 550–600 °C. The combustion performance index (*S*) of the raw (B and SCF) and beneficiated coal samples (oxidized) were also determined. It is observed that the beneficiated SCF coal samples have lesser combustion performance index than that of the raw coal samples (Table 2). However, the B coal samples shows higher combustion performance index than that of raw coal samples. Thus, the beneficiated B coal samples will have

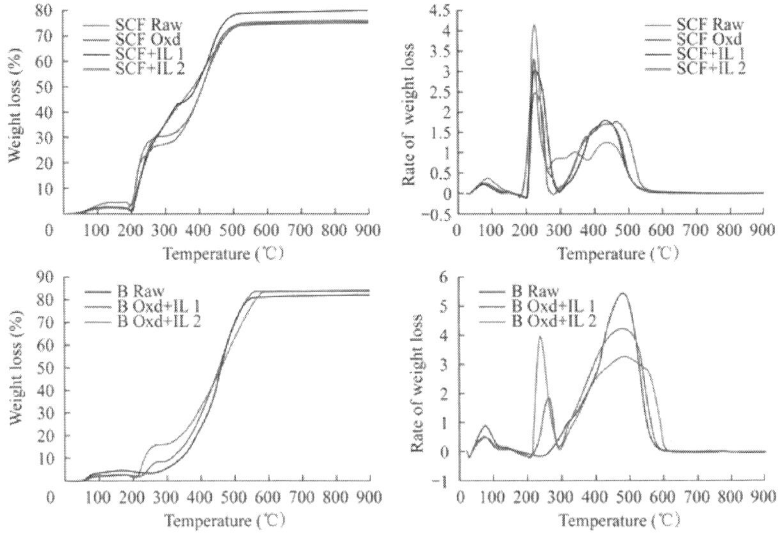

Figure 8. TG–DTG of raw and beneficiated (B and SCF) coal samples

Table 2. Combustion characteristics of coals

Coal samples	Ignition temperature (T_i)	Burnout temperature (T_b)	The combustion performance index ($S \times 10^{-7}$)	Activation energy (Ea; kJ/mol)
SCF	222.18	586.53	3.18	164.64
SCF Oxd	240.13	564.85	1.25	155.34
SCF + IL1	219.13	564.85	1.40	192.90
SCF + IL2	219.13	546.79	1.22	192.93
B	483.61	625.55	356.8	208.00
B Oxd	492.47	641.91	323.0	241.30
B Oxd + IL1	485.74	627.43	418.0	163.70
B Oxd + IL2	487.91	652.67	450.8	181.42

higher combustion activity in the thermal application at lower temperatures. The combustion performance indexes (*S*) along with the activation energies (*E*a) of the raw and beneficiated coal samples are listed in the Table 2, which shows significant variations with different types of the treatments. The activation energies (*E*a) of the beneficiated coals are also found to be generally lesser than the raw coals that decrease with the rise in their S values. However, the activation energies of the IL oxidized coals are found to be higher, which may be due to the increase in carbon contents during the oxidation. It is to be noted that the lower the ignition temperature the easier the ignition of coal. The lower the burnout temperature, the shorter the burnout time of the coal, and the lower the combustible matter content of the ash. Thus, it is obvious that the initial ignition and burnout temperatures of coal samples not only vary with the reactants but also with the coal properties. The present partially green process can also be adopted for the development of value added products from coals.

4 CONCLUSIONS

The ILs can be used for the beneficiation of high sulfur coals through removal all forms of sulfur including mineral matters. In the present process, ILs are able to reduce the pyritic, sulfate and organic sulfur by maximum of 62.50 %, 83.33 % and 31.63 % from Meghalya (India) coals. The FT-IR, XRF and SEM analyses support the beneficiation process through removal of sulfur components and mineral matters from the coal samples. The GCVs of the beneficiated coals are enhanced after oxidation in ILs. The process removed some of inorganic species such as Ca, Mn and Ni. The combustion performance indexes (*S*) and activation energies (*E*a) indicate the suitability of the beneficiated coals for thermal applications. Moreover, the chemical process finds merits for industrial application as it could be carried out in mild conditions. In summary, the present green process of chemical desulfurization of coals with ILs is observed to substitute the use lethal chemicals in coal beneficiation.

ACKNOWLEDGMENTS

The funding from the Ministry of Steel, Govt. of India (GAP-249) and CSIR (MLP-6000-WP-III) are highly acknowledged. Authors express thanks to the USIC-NEHU, Shillong (India) for the SEM images. Dr. Yongfeng Hu and Dr. Stephen G. Urquhart are highly acknowledged for the XRF spectra of coal samples at the Canadian Light Source, University of Saskatchewan. Many thanks are also given to Dr. Robert Finkelman for his comments to the draft.

REFERENCES

1. Ahmed M, Rahim A (1996) Chemical desulfurization methods of organic sulfur of Eocene coal, Jaintla Hills, Meghalaya. In: Fuel and Energy Abstracts vol 5. p 333
2. Ahmed A, Ahmad N, Shah R, Bhatti MN, Saleem M (2008) Coal desulfurization by solvent leaching methods. J Fac Eng Technol 15:47–56
3. ASTM (1991) Annual book of ASTM standards; coal and coke. Standard practice for proximate analysis of coal and coke 05:05
4. ASTM (2005) Annual book of ASTM standards; coal and coke. Standard test method for forms of sulfur in coal 05:06
5. ASTM (2010) Standard practice for collection of a gross sample of coal. ASTM International, West Conshohocken **2011**
6. ASTM (2013) Annual book of ASTM standards; gaseous fuels; coal and coke; bioenergy and industrial chemicals from biomass. Standard test method for gross calorific value of coal and coke 05:06
7. Barooah PK, Baruah MK (1996) Sulphur in Assam coal. Fuel Process Technol 46:83–88
8. Baruah BP, Khare P (2007) Desulfurization of oxidized indian coals with solvent extraction and alkali treatment. Energy Fuels 21:2156–2164
9. Baruah BP, Saikia BK, Kotoky P, Rao PG (2006) Aqueous leaching on high sulfur sub-bituminous coals, in Assam, India. Energy Fuels 20:1550–1555
10. Bosmann A, Datsevich L, Jess A, Lauter A, Schmitz C, Wasserscheid P (2001) Deep desulfurization of diesel fuel by extraction with ionic liquids. Chem Commun 23:2494–2495
11. Calkins WH (1994) The chemical forms of sulfur in coal: a review. Fuel 73:475–484

12. Campos-Martin JM, Capel-Sanchez MC, Perez-Presas P, Fierro JLG (2010) Oxidative processes of desulfurization of liquid fuels. J Chem Tech Biotech 85:879–890

13. Chou CL (2012) Sulfur in coals: a review of geochemistry and origins. Int J Coal Geol 100:1–13

14. Davalos A, Pecina ET, Soria M, Carrillo FR (1994) Kinetics of coal desulfurization in an oxidative acid media. Int J Coal Prep Util 29:152–172

15. De Filippis P, Scarsella M (2003) Oxidative desulfurization: oxidation reactivity of sulfur compounds in different organic matrixes. Energy Fuel 17:1452–1455

16. Demirbas A, Balat M (2004) Coal desulfurization via different methods. Energy Sour 26:541–550

17. Ehsani MR (2006) Desulfurization of tabas coals using chemical reagents. Iran J Chem Eng 25:59–66

18. Esser J, Wasserscheid P, Jess A (2004) Deep desulfurization of oil refinery streams by extraction with ionic liquids. Green Chem 6:316–322

19. Falcon R, Ham AJ (1988) The characteristics of Southern African Coals. J S Afr Inst Min Metall 88:145–161

20. He L, Li H, Zhu W, Guo J, Jiang XLu (2008) Deep oxidative desulfurization of fuels using peroxophosphomolybdate catalysts in ionic liquids. Ind Eng Chem Res 47:6890–6895

21. Holbrey JD, Lopez-Martin I, Rothenberg G, Seddon KR, Silvero G, Zheng X (2008) Desulfurisation of oils using ionic liquids: selection of cationic and anionic components to enhance extraction efficiency. Green Chem 10:87–92

22. Ishak MAM, Ismail KB (2004) Sulfur removal from Mae Moh coal via mixture of hydrogen peroxide and formic acid. In: Proceedings of the 21st international Pittsburgh coal conference

23. Juniper LA, Pohl JU (1997) What price coal quality in electric power generation? The Australian Coal Review (7):37–42

24. Kendra-Krolik K, Fabrice M, Jaubert JN (2011) Extraction of thiophene or pyridine from n-heptane using ionic liquids gasoline and diesel desulfurization. Ind Eng Chem Res 50:2296–2306

25. Komintarachat C, Trakarnpruk W (2006) Oxidative desulfurization using polyoxometalates. Ind Eng Chem Res 45:1853–1856

26. Maes II, Gryglewicz G, Machnikowska H, Yperman J, Franco DV, Mullens J (1997) Rank dependence of organic sulfur functionalities in coal. Fuel 76:391–396

27. Mazumdar B, Saikia PC, Sain B, Baruah BP, Bordoloi CS (1989) Desulphurisation of high sulphur coals of north-east Indian Coalfields by organometallic compounds. Fuel 68:610–613

28. McCoy RN, Weiss FT (1954) Application of catalytic desulfurization to group sulfur analysis of petroleum fractions. Anal Chem 26:1928–1933

29. Nayak B, Singh AK, Upadhyay AK, September Bhattacharyya KK (2009) A note on the characters of some lower Gondwana coals of West Siang district in the Arunachal Himalaya and their trace element content. J Geol Soc India 74:395–401

30. Qing W, Hao X, Hongpeng L, Chunxia J, Jingru B (2011) Thermogravimetric analysis of the combustion characteristics of oil shale semi-coke/biomass blends. Oil Shale 28:284–295

31. Rogers RD, Seddon KR (Eds) (2003) Ionic liquids as green solvents: progress and prospects. ACS Symposium Series

32. Saikia BK, Boruah RK, Gogoi PK, Baruah BP (2009) A thermal investigation on coals from Assam (India). Fuel Process Technol 90:196–203

33. Saikia BK, Ward CR, Oliveira MLS, Hower JC, Baruah BP, Braga M, Silva LF (2014) Geochemistry and nano-mineralogy of two medium-sulfur northeast Indian coals. Int J Coal Geol 121:26–34

34. Sain B, Saikia PC, Baruah BP, Bordoloi CS, Mazumder B (1991) Desulphurization of Assam coal by chlorinolysis and hydrothermal techniques. Fuel 70:753–755

35. Schmitz C, Datsevich L, Jess A (2004) Deep desulfurization of diesel oil: kinetic studies and process-improvement by the use of a two-phase reactor with pre-saturator. Chem Eng Sci 59:2821–2829

36. Silva G, Voth S, Szymanski P, Prokopchuk EM (2011) Oxidation of dibenzothiophene by hydrogen peroxide in the presence of bis(acetylacetonato)oxovanadium(IV). Fuel Process Technol 92:1656–1661

37. Silverstein RM, Bassler GC, Morril TC (1987) Spectrometric identification of organic compunds, 5th edn. Wiley, New York

38. Sonibare OO, Ehinola OA, Egashira R, KeanGiap L (2005) An investigation into the thermal decomposition of Nigerian coal. J Appl Sci 5:104–107

39. Stracher GB, Taylor TP (2004) Coal fires burning out of control around the world: thermodynamic recipe for environmental catastrophe. Int J Coal Geol 59:7–17

40. Yaman S, Ersoy-Mericboyu A, Küçükbayrak S (1954) Chemical coal desulfurization research in turkey. Fuel Sci Technol Int 13:49–58

41. Yperman J, Franco D, Mullens J, Van Poucke LC, Gryglewicz G, Jasienko S (1995) Determination of sulfur groups in pyrolysed low-rank coal by atmospheric-pressure T.P.R. Fuel 74:1261–1266

42. Zhang G, Yu F, Wang R (2009) Research advances in oxidative desulfurization technologies for the production of low sulfur fuel oils. Pet Coal 51:196–207

43. Zhao H, Xia S, Ma P (2005) Use of ionic liquids as 'green' solvents for extractions. J Chem Technol Biotechnol 80:1089–1096

44. Zhao D, Wang Y, Duan E, Zhang J (2010) Oxidation desulfurization of fuel using pyridinium-based ionic liquids as phase-transfer catalysts. Fuel Process Technol 91:1803–1806

6

CHAPTER

A Novel Suspension-Floating-Circulating Fluidized Combustion Technology for Coal Slurry

Yi Wang[1] • Maoyong Cao[1] • Zenghui Wang[2] • Lu Wei[1] • Shuping Zhao[3]

[1] *Shandong University of Science and Technology, Qingdao 266590, China*

[2] *Department of Electrical and Mining Engineering, University of South Africa, Pretoria 1710, Florida, South Africa*

[3] *Shandong Energy Feicheng Mining Group Co. LTD Baizhuang Mine, Feicheng 271600, China*

ABSTRACT

A novel suspension-floating-circulating fluidized combustion technology is proposed for burning coal slurry fuel in traditional circulating fluidized bed boilers (CFBB). This technology can solve some existing problems in large-capacity CFBB burning coal slurry. The principles of the suspension-floating-circulating fluidized combustion technology were introduced in detail in this paper. A 130 t/h CFBB was retrofitted based on the technology, and the retrofitted system mainly includes a long-distance transport sub-system, a bed-material conveying sub-system with a wind-seal device invented by the authors, a superheater thermoregulation device using a novel temperature regulator, a return loop fluidization facility, and a pneumatic ash conveying sub-system with sealed pump. The achieved performance of the retrofitted CFBB shows that the thermal efficiency is 89.83 %, the combustion efficiency is 96.24 %, and the blending proportion of slurry is 94 %.

Keywords: Suspension-floating-circulating fluidized Coal slurry Combustion technology CFBB

1 INTRODUCTION

Coal slurry is a by-production of coal washing plant. The surface of slurry is viscous with the water content of 25 to 40 percent and 80 percent of solid coal particles are less than 3 mm in diameter (Cao et al. 2009; Liu et al. 2012), so it is difficult to achieve long distance carriage. As a result, a lot of coal slurry has been stacking in the air for a long time, they not only encroach a great quantity of land but also cause serious damage to ecological environment.

Extensive utilization of low-heat value fuel is important in China. Coal slurry belongs to the low-heat value fuel, whose combustion technology has been studied extensively in recent years (Chi et al. 2001; Tang et al. 2007; Wu et al. 2009; Bo et al. 2013; Ni et al. 2014; Shu and Zhang 2014). The results reveal that appropriately retrofitted circulating fluidized bed (CFB) boilers of less than 75 t/h capacity can be successfully applied in burning coal slurry, but it is still difficult to solve the problem of coal slurry large-scale application. So researches are paying increasing attention to CFBB with capacities larger than 75 t/h. While most of the previous researches only focused on the atomization combustion (Jakobs et al.2012; Daviault et al. 2012; Ma et al. 2013). It was reported that a retrofitted 130 t/h CFBB with atomization combustion had achieved boiler thermal efficiency of 90.3 % and combustion efficiency of 99 % (Cheng et al. 2008). A retrofitted 440 t/h CFBB with atomization combustion (Yu et al. 2008; Qu et al. 2010) has achieved a boiler thermal efficiency of 89 %, but it only blended 25 % coal slurry. From these applications, it can be concluded that large-scale CFB boilers have difficulties in achieving high boiler thermal efficiency, combustion efficiency and blending proportion in burning coal slurry. The coal slurry atomization combustion has some problems as follows: (1) coal slurry needs special handling, (2) the high-pressure atomization devices are expensive and easy to be blocked, (3) the life of the feeding hose is short, (4) slagging easily appear at the bottom of the fluidized bed, (5) fly ash carbon content is high, and (6) boiler efficiency is low, etc.

The combustion mode of the feeding slurry from the top of the boiler has been also studied in recent years (Gagarin and Gyu'maliev 2009; Wang et al. 2010), and it is considered to have good prospect for development. A 75 t/h CFB boiler with the combustion have achieved a boiler thermal efficiency of 86 % and an average combustion efficiency of 95 % after retrofitted by blending 70 % coal slurry (Ji and Wang2007). However, the application results showed that there were still some problems that need to be solved:

1. The blending proportion of slurry is usually less than 70 %;
2. The over-temperature phenomenon of platen superheater is serious causes explosion easily;
3. The temperature of returned fuel is high which can easily cause coking in the return feeder;
4. The high temperature ash goes into the coal conduit easily that makes the conveying belt aging;
5. The contentration of sulfur and nitrogen in the fly ash is high;
6. The loss and failure rate of the ash conveying system are high.

These problems often lead to the lower boiler thermal efficiency, combustion efficiency and blending proportion in burning coal slurry, therefore preventing the suspension combustion technology from being applied to the larger CFB boiler.

Based on the large-scale CFB boilers combustion technology especially on the existing problems, a new type of suspension-floating-circulating fluidized combustion technology is necessary and presented in this paper. The main principles and design concepts of the technology are introduced in detail in the following section, and the results of the practical application of a 130 t/h CFBB, which is used in a power plant, are presented. Finally, the conclusion is given.

2 SUSPENSION-FLOATING-CIRCULATING FLUIDIZED COMBUSTION TECHNOLOGY

High concentration coal slurry is used as boiler fuel. Coal gangue is used as bed material. The structure of the boiler system using suspension-floating-circulating fluidized combustion technology is shown in Fig. 1.

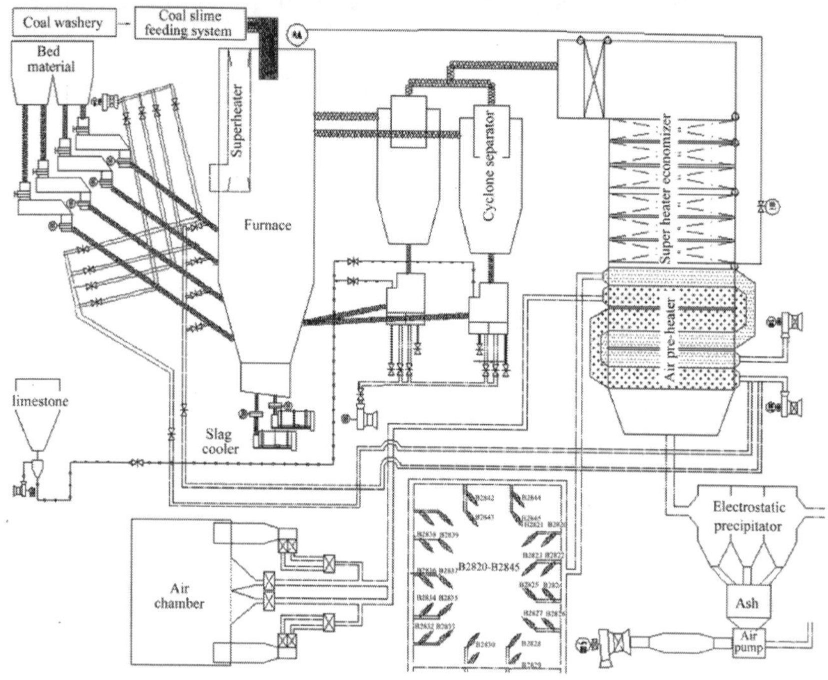

Figure 1. Structure of the boiler system

The boiler mainly includes the furnace, coal slurry feeding system, the bed-material conveying system, steam temperature regulated system, a cyclone separator and an U return feeder, an ash conveying system, a primary air and secondary air system and bottom slag cooler, etc. Compared with atomization combustion technology, slurry atomization, filtering and high pressure feeding devices are not needed, so the boiler using the new technology, which doesn't require extensive retrofit and the reliability can be greatly improved.

As shown in Fig. 2, the principles of the new technology are as follows. Some of the high density and different-size coal gangue granules are conveyed to the bottom of the fluidized bed by the bed-material conveying system to ensure the bed pressure of the CFB. The high concentration slurry from the coal washery with a moisture content of about 30 % is delivered to the top of the furnace by a feeding system, and fallen into the furnace along the slurry pipeline. The falling coal slurry meet the high temperature (900–960 °C) of furnace and experience drying, agglomeration, volatilize, thermal explosion, burning and falling stages. This process is the suspension combustion. Through the suspension combustion, big slurry blocks break into different-size blocks. Asymmetrical secondary air is designed to flow on the opposite sides of the furnace wall. Under the influence, the falling coal slurry blocks rotate, thus its falling time and path are prolonged to make the blocks finish suspension combustion. After the suspension combustion, the unburned blocks fall on the CFB surface, float above the coal gangue, and finish the floating combustion. With the bed material fluidization period, some unburned blocks will experience fragmenting, abrasion, crushing, aggregation, rising and sinking, blowing away and burn out. This is the internal circulating fluidized combustion process. The other parts of unburned blocks and the flue gas enter the cyclone separator for gas–solid separation together. When the separated gas enters the rear smoke channel, the separated ash will be transported to the ash warehouse after the electrostatic precipitator. While the separated solid including the bed material and unburned blocks will return to the CFB via the return feeder to go to the next combustion cycle. This process is the external circulating fluidized combustion. The coal slurry experiences suspension-floating-circulating fluidized combustion, and eventually achieves efficient combustion.

In order to ensure the combustion efficiency, and improve the blending proportion of slurry and the reliability of the boiler, the CFBB's auxiliary system has been designed. It mainly includes a long distance feeding system with plunger pump, a bed-material conveying system with the new sealing device, a three levels of superheater thermoregulation device with two first-grade desuperheaters, a cyclone separator and U return feeder, a low pressure pneumatic ash conveying system with material sealed pump and an asymmetrical secondary air system on both sides of the furnace.

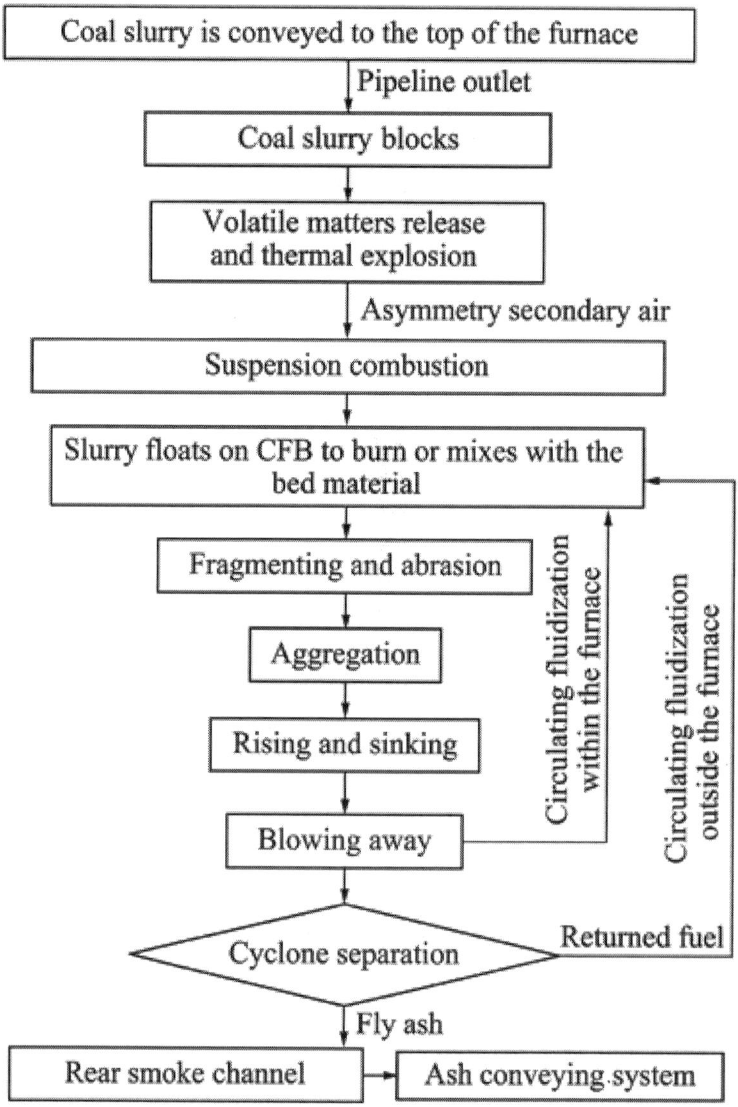

Figure 2. Combustion process flowchart

The new technology has the following features. The blending proportion of coal slurry can reach 94 % in the CFBB fuel; the coal slurry is fed to the top of furnace via pipeline directly, which eliminates the need of atomization and filtering systems in atomization combustion; and the high combustion temperature (900–960 °C) can improve the burnout effect significantly and effectively control the coking and N_2O production; the new combustion technology can also solve many problems, such as the superheater over temperature, the conveying belt of bed material aging, high content of sulfur and nitrogen in ash, and returned fuel coking in the return feeder.

2.1 Coal slurry feeding system

Pipeline is the main method for conveying the coal slurry (Chen et al. 2009; Liu and Duan 2009). Long-distance pipeline transportation is the key of the coal slurry burning. The viscous material pipeline system (MNS pipeline transportation system) (Lu and Duan 2006) is used to convey the slurry. It could solve a series of technical problems in the long-distance pipeline transport, such as the slurry flow control, duct sealing vibration, pipes wear, pulping, storing, mixing, transporting, feeding and cleaning. It can achieve a transported process which is fully enclosed, steady, long-distance and non-polluting.

The structure of MNS is shown in Fig. 3. In the MNS, slurry is sent to the cache warehouse after pretreatment, then through overhaul gate, positive pressure feeder, double plunger pump, valve, and finally fed into the furnace via spiral feeder. The work ing process is controlled automatically by a programmable logic controller and a distributed control system.

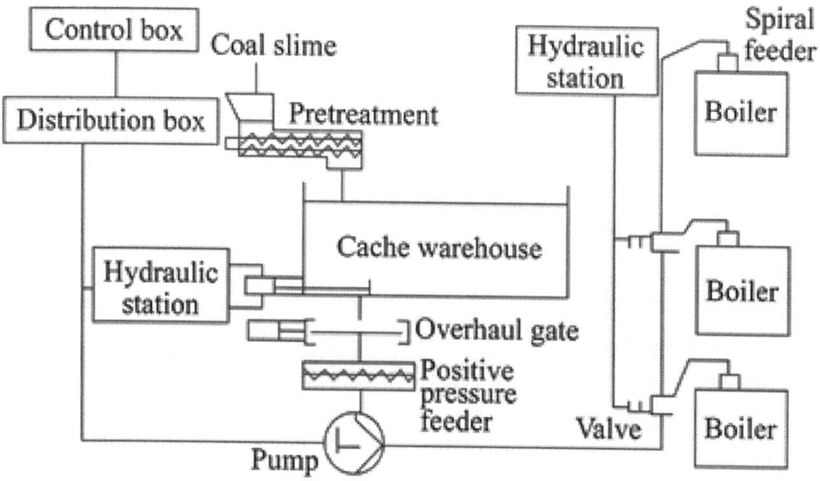

Figure 3. MNS pipeline transportation process flowchart

The diameter of pipeline and the average velocity of coal slurry can be calculated by the formulas.

$$\overline{V} < \sqrt[2-n]{R_e 8^{n-1} \rho^{-1} \eta D^n}$$

$$\overline{V} = \frac{4Q}{\pi D^2}$$

where, R_e is the Reynolds number, Q is the desired flow of the CFBB, ρ is the density of coal slurry, n and η are rheological characteristic parameters, and D is the internal diameter of the pipeline.

2.2 Bed-material conveying system

In the process of boiler start-up and operation, keeping a steady bed material height is very important for boiler safety. The bed material conveying system is designed based on the original coal conveying system of CFB boiler. Because the particles density and distribution size of bed material are higher than the coal particles, it easily causes blockage of the pipeline which used to transport the bed material.

Moreover, when no bed material in the conveying pipe, the high temperature flue gas in the furnace returns enters to the pipe easily and then causes the conveying belt aging.

For these reasons, a new air sealing device is designed (Fig. 4). Two symmetrical sealing air pipes are set at the bottom of the conveying pipe. If the pressures of the sealing air are F_1 and F_2, then F_1 and F_2form a joint force F_3. F_3 forms a resistance to prevent the high temperature flue gas. At the same time, the pressure of conveying air and F_3 form the push–pull effect in conveying pipe to solve the problem of pipe blockage.

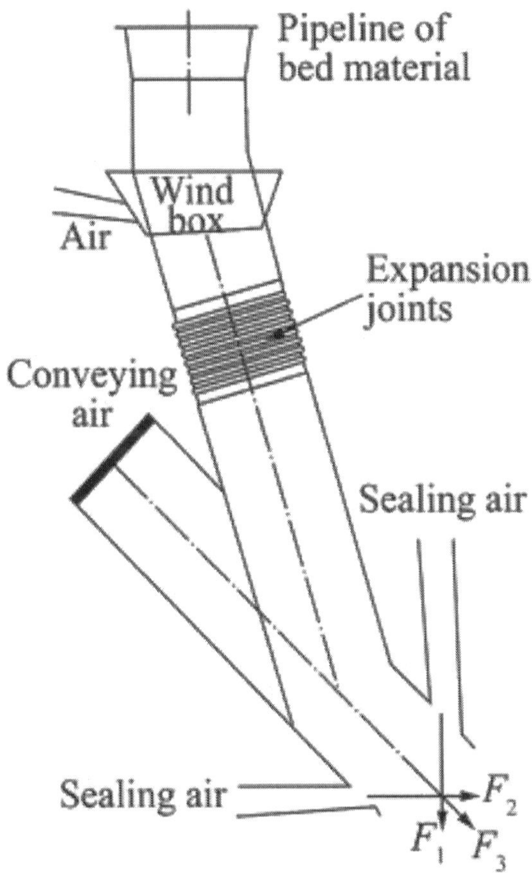

Figure 4. Bed material conveying pipe structure

The bed-material conveying system has the following merits. It can convey widely size bed material. The innovative sealed device can not only decrease the pipe blockage, but also prevent the high-temperature flue gas to prolong the life of the conveying belt.

2.3 Furnace system design

The furnace is the most important component in the suspension-floating-circulating fluidized combustion system. Materials enter into the furnace include fuel, bed material, desulfurizer and gas, etc. Meanwhile some solid materials and gases leave the furnace, including the ash at the bottom of furnace, recycled material and flue gas.

2.3.1 Heat exchange law inside the furnace

The heat exchange inside the furnace satisfies the following equations (Jiang and Xiao 2009).

$$B_j Q_1^R = \phi B_j [X Q_{1s}^m - (I''_{mz} - I''_{ze}) \frac{B_s}{B_j} - I''_m]$$

$$B_j Q_1^R = K_m B_j^m (\theta''_m - t_b)$$

$$Q_{1s}^m = Q_{net,ar}\eta + \alpha'_m I'_{rk} + \alpha'_z I'_{lk} + \alpha'_p I'_{lk}$$

where, B_j is the fuel consumption, kg/s; QR1Q1R is the heat of per kilogram of combustion product transferred to the heating surfaces and material in the furnace, kJ; ϕ is the coefficient of heat preservation; Qm1sQ1sm is the effective value of heat that is released, kJ/kg; x is the share of effective value Qm1sQ1sm; I″mzImz″ is the ash enthalpy at furnace exit, kJ/kg; I″zeIze″ is the ash enthalpy at furnace inlet, kJ/kg; B_s is the amount of circulating ash, kg/s; I″mIm″ is flue gas enthalpy at furnace exit, kJ/kg; k_m is the heat transfer coefficient of furnace, kJ/(m^2 s °C); BmjBjm is the heating surface area of furnace, m^2; θ″mθm″ is the temperature at furnace exit, °C; t_b is the tube wall temperature at the water wall, °C; α′mαm′ is the excess air coefficient at furnace exit, kJ/kg; α′zαz′ is the excess air coefficient at return valve inlet; I′lkIlk′ is the cold air enthalpy, kJ/kg.

The equation of boiler thermal efficiency is as follows (Jiang and Xiao 2009).

$$\gamma = 100 - \frac{Q_2 + Q_3 + Q_4 + Q_5 + Q_6 + Q_7}{Q_r} \times 100\%$$

where, γ is the boiler thermal efficiency; Q_2 is the exhaust smoke heat loss per kilogram fuel, kJ; Q_3 is the combustion heat loss of combustible gas per kilogram fuel, kJ; Q_4 is the solid incomplete combustion heat loss per kilogram fuel, kJ; Q_5 is the heat loss per kilogram fuel, kJ; Q_6 is the slag heat loss per kilogram fuel, kJ; Q_7 is the desulfurization heat loss per kilogram fuel, kJ; Q_γ is the heat input per kilogram fuel, kJ.

2.3.2 Key parameters of the furnace

Bed temperature is one of key parameters of the suspension-floating-circulating fluidized combustion system. The designed temperature should avoid grey softening and coking, improve the combustion efficiency, and reduce the emissions of fly ash carbon and N_2O. In addition, the slurry is difficult to burn and volatilize, so the bed temperature is designed to be 900–960 °C.

The furnace heat transfer coefficient (Jiang and Xiao 2009) is estimated by this formula.

$$k = k_0 \alpha^A T^B$$

where, k is the heat transfer coefficient, kW/m^2; k_0 is the ratio; A and B are constant coefficients; α is the concentration of suspended particles, kg/m^3; and T is the average temperature of the furnace. The material concentration is the biggest factor affecting heat transfer coefficient of the heating surface of the indoor combustion, higher materials concentration makes higher heat transfer coefficient.

Correct choice of the furnace fluidization air speed is also important to slurry combustion. The fluidization air speed is determined generally by fluidized bed cold experiment.

2.3.3 Secondary air system design

If the fuel and air is not mixed evenly, it can affect the thermal efficiency and the combustion efficiency of the boiler. In the furnace combustion center of the traditional CFBB, there is an oxygen-depleted zone (Jiang and Xiao 2009), as shown in Fig. 5. The oxygen-depleted zone is very detrimental to slurry suspension combustion. Under the premise of guaranteed boiler pressure, as shown in Fig. 6, the multilevel asymmetry design of secondary air can carry oxygen to the furnace during the slurry blocks fall down in the process of S-shaped rotation. This design can avoid the oxygen-depleted zone and extend the falling time and path of slime to ensure the combustion efficiency and desulfurization efficiency (Gunka and Pyshyev 2014).

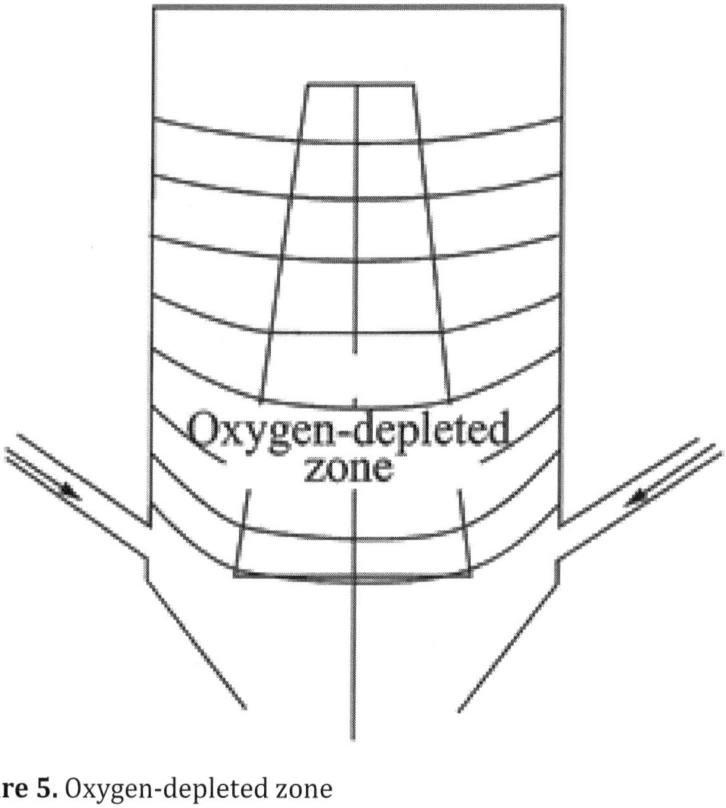

Figure 5. Oxygen-depleted zone

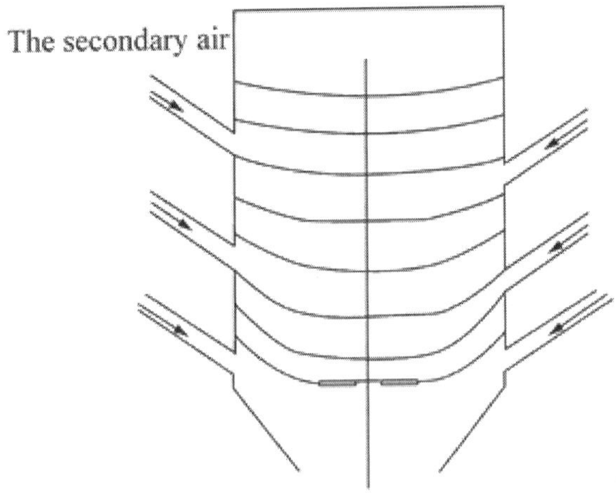

Figure 6. Asymmetry secondary air

2.4 Steam temperature regulated system

The steam temperature of the boiler will affect the safety and economy of the whole production process directly. Steam temperature regulated system of the boiler is composed of a three-level superheater and two-level desuperheater.

When slurry is burning in the furnace, the suspension combustion will make the superheater system deviated from its nominal parameters, especially the over-temperature phenomena of the platen superheater is a serious problem to boiler safty. For this reason, a new types of spray desuperheaters is devised, as shown in Fig. 7.

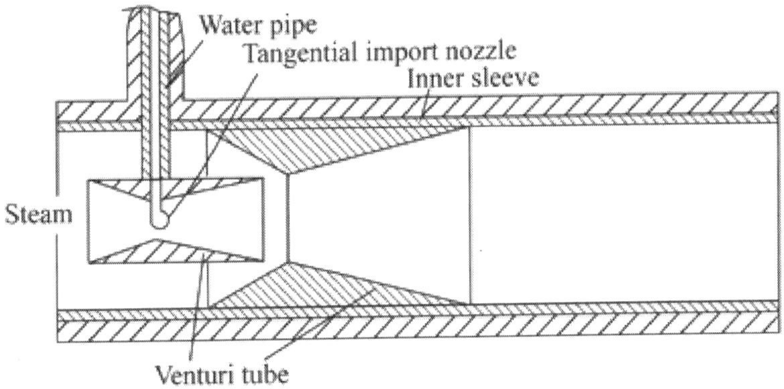

Figure 7. The structure of new desuperheater

It consists of a steam pipe, water pipe, venturi tube, tangential import centrifugal nozzle and inner sleeve. The new type desuperheater is set before the platen superheater. It has well atomization effect and high thermal efficiency; it can realize a wide range of steam temperature regulation; and it's able to adapt to a wide range of steam loads.

2.5 Fuel return system

The system can improve the fuel combustion efficiency and reduce the thermal heat losses of solid fuel. It mainly includes a cyclone separator, U return feeder, and fluidizing air, which are critical components to the fuel circulating. During the suspension combustion, the high-temperature fuel gas flies out of the furnace and enter the cyclone separator. The unburned fuel particles will be separated and drop into the U feeder, finally under the effect of fluidizing air they return to the furnace. If general compressed air is used to fluidize the returned fuel, the oxygen in the compressed air will easily cause the returned high-temperature fuel to burn and coke in the feeder, and that will affect the safe operation of the boiler. So steam is used to replace compressed air in the new technology (Li and Xiao 2005). The steam velocity V_{lj}can be calculated as follows.

$$V_{\mathrm{lj}} = 0.0882A_{\mathrm{r}}^{0.528}\left(v_{\mathrm{y}}/D_{\mathrm{dl}}\right)\ \mathrm{m/s}$$

where, $A_{\mathrm{r}} = gD_{\mathrm{dl}}^{3}/v_{\mathrm{y}}^{2}('y_{\mathrm{g}} - 'y_{\mathrm{y}})/'y_{\mathrm{y}}$ is the rule of Archimedes; v_y is the coefficient of kinematic viscosity of steam, m/s^2; $'y_g$ is the true specific gravity of solid particle, kg/m^3; $'y_y$ is the portion of steam, kg/m^3; D_{dl} is the average diameter of the solid particle, m; g is the acceleration of gravity, m/s^2.

The limestone ($CaCO_3$) as desulfurizer under the high temperature will decompose as follows:

$$\mathrm{CaCO_3} \rightarrow \mathrm{CaO} + \mathrm{CO_2}$$

Then the generated CaO meet the H_2O react in the steam as follows:

$$CaO + H_2O \rightarrow Ca(OH)_2$$

Under the same Ca/S molar ratio, the desulfurization efficiency of $Ca(OH)_2$ is higher than that of $CaCO_3$. Therefore using steam instead of compressed air cannot only avoid the secondary combustion of the returned fuel but also increase significantly the desulfurization efficiency.

2.6 Ash conveying system

After the slurry burn-out, a large amount of fine ash will be produced. If the ash is not discharged in time, it will affect the electrostatic precipitator and pollute the environment (Wang et al. 2014). In many CFBB bins pumps are used to convey ash. It shows large power consumption, serious wear and tear. To avoid these problems, the LFB pump with low pressure air is chosen in this new system, and its structure is shown in Fig. 8.

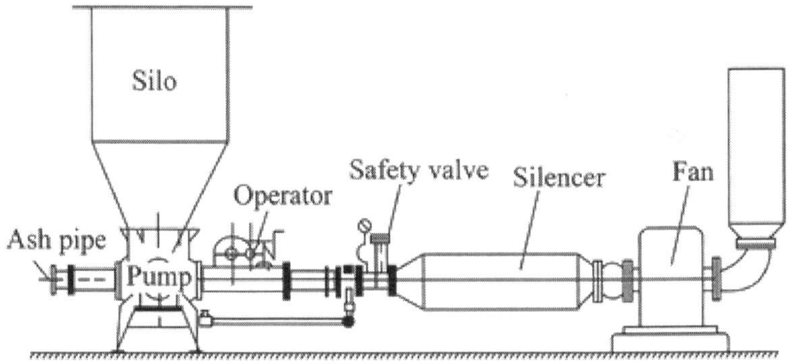

Figure 8. The structure of material sealed pump

The energy of the LFB pump comes from the fan and the potential energy of the silo. Low pressure air goes through the duct and pump into the diffusion chamber. When high speed air goes through the pump, it forms negative pressure between the nozzle and the diffusion chamber, the gasified ash by nozzle is sucked into the ash pipe.

Table 1. Design parameters of the boiler

No.	Name	Units	Value
1	Rated steam temperature	°C	540
2	Rated steam pressure	MPa	9.8
3	Rated capacity	t/h	130
4	Maximum evaporation	t/h	143
5	Feed-water temperature	°C	215
6	Boiler flue gas temperature	°C	143
7	Thermal efficiency	%	89.58
8	Coal slurry consumption	kg/h	28,528
9	Limestone consumption	kg/s	1.24
10	Limestone particle size	mm	≤2
11	Hot air temperature	°C	300
12	Cold air temperature	°C	30
13	Flow ratio of pri. and 2nd air		50:50
14	Superheater steam temperature	°C	520
15	Combustion chamber temperature	°C	960
16	Blowdown rate	%	≤2
17	Circulation ratio		15–20
18	Desulphurization efficiency	%	≥85
19	Calcium sulfur ratio		1.5–2
20	Fly ash of the boiler	kg/s	8.4
21	Design boiler efficiency	%	93.24

3 130 T/H SUSPENSION-FLOATING-CIRCULATING FLUIDIZED BED BOILER

3.1 The design parameters of the boiler

The type of CFBB used in the paper is TG130/9.8-M. It is characterized by high temperature and pressure, single drum tap, single furnace, natural circulation, and full steel frame layout. It is equipped with two sets of 30 MW steam turbine generator. Its original fuel is bituminous coal or anthracite. The boiler is retrofitted to burn slurry, and its main design parameters are shown in Table 1.

3.2 The fuel of boiler

The parameters of the desired fuel for the boiler are shown in Table 2.

Table 2. Proximate and elementary analysis of coal slurry

Item	Proximate analysis				$Q_{net,ar}$ (kJ/kg)	Elementary analysis				
	M_{ad} (%)	A_{ad} (%)	V_{ad} (%)	FC_{ad} (%)		C_{ar} (%)	H_{ar} (%)	N_{ar} (%)	$S_{t,ar}$ (%)	O_{ar} (%)
Design slurry	34	18.35	19.36	32.16	17,800	39.28	2.60	0.71	0.27	5.14
Tested slurry	35.33	11.53	20.25	34.67	18,423	43.20	2.53	0.95	0.38	3.35

The coal slurry with a moisture content of 25 %–30 % and size of particles of less than 3 mm from the Baizhuang mine is used to show the new technology. The test data of Baizhuang coal slurry is also shown in Table 2. Detailed analytical data of the design and test of coal slurry shows that the coal slurry from the Baizhuang mine can be used as the boiler fuel. The blending proportion of slurry is 94 % in boiler fuel.

3.3 **Selection of bed material**

The selection of density and size distribution of inert bed material directly affects the fluidization quality of the boiler, and further affects the characteristics of combustion and heat transfer. Considering the different density fluidized bed and operating cost, the coal gangue in Baizhang mine is selected as the bed material. The proximate analysis of this coal gangue is shown in Table 3.

Table 3. Proximate analysis of coal gangue

Item	M_{ad} (%)	A_{ad} (%)	$S_{t,ar}$ (%)	$Q_{net,ar}$ (MJ/kg)	FC_{ad} (%)
Gangue	3	80	0.15	5.3	17

Coal gangue is waste in coal preparation process and washing coal, the diameter of particles is generally 0–10 mm, the density is 1.69 g/cm³, and deformation temperature is 1350 °C. The size distribution of fresh bed material is shown in Table 4.

Table 4. Size distribution of bed material

No.	Aperture of screen (mm)	Average diameter(mm)	Percentage (%)
1	2–4	2	5
2	4–6	4	30
3	6–8	6	60
4	8–10	8	5

Using coal gangue as bed material has the following advantages: firstly, the hardness of coal gangue is relatively low, so it can effectively reduce the wear and tear of the boiler; secondly, after coal gangue is added as bed material, some particles will participate in the combustion to generate heat, so as to improve the combustion efficiency of the boiler; thirdly, it can make full use of waste materials.

3.4 Coal slurry conveying system

As shown in Fig. 9, the coal slurry conveying system of the 130 t/h boiler is used to transport slurry with concentration of 70 %–75 % from the Baizhuang mine. The transport distance is 300 m in the boiler system.

Figure 9. Coal slurry conveying equipment

According to the rheological properties of slurry experiment, the reynolds number R_e is 2100, the desired flow of the boiler Q is 7.92 kg/s, the density of coal slurry ρ is 1430 kg/m^3, the rheological index n approaches to 1, the power-law viscosity coefficient η is 0.2011, the maximum pressure of the hydraulic system is set to 31.5 MPa, and the pipeline pressure is set to 18 MPa. Then the conveying pipeline diameter D is 291 mm, the maximum flow rate of the system is set to 30 m^3/h, and the maximum conveying distance is longer than 500 m.

3.5 Bed-material conveying system

The bed-material conveying system guaranteeing the layer height and pressure of the coal gangue, makes the boiler maintain a good circulating fluidized state and effectively avoids coking after slurry falling on the CFB, and then ensures the safe operation of the boiler.

The cold test of CFB shows that the layer height of the coal gangue on the CFB should be between 550 mm and 600 mm, and the differential pressure of the gangue layers should be between 6.8 kPa and 7.2 kPa. A bed-material conveying system with a wind-seal device invented by the authors is designed to convey bed material safely, as shown in Fig. 10, sealed pipe is 100 mm in diameter, sealed wind flow is about 1300 m^3/min, sealed air pressure is not less than 7.5 kPa.

Figure 10. Wind-seal equipment

3.6 Steam temperature regulated system

In the process of boiler operation, the phenomenon of overtemperature at the platen superheater is very serious in the steam temperature regulated system. The safe inlet temperature of platen superheater must be less than 400 °C. The platen superheater temperature is greatly influenced by the flue gas temperature at furnace exit. The flue gas temperature at furnace exit is mainly affected by the blending proportion of slurry. The relation between the flue gas temperature at furnace exit and the blending proportion of slurry is shown in Fig. 11. So the higher the blending proportion of slurry is, the bigger the temperature value of the platen superheater will be.

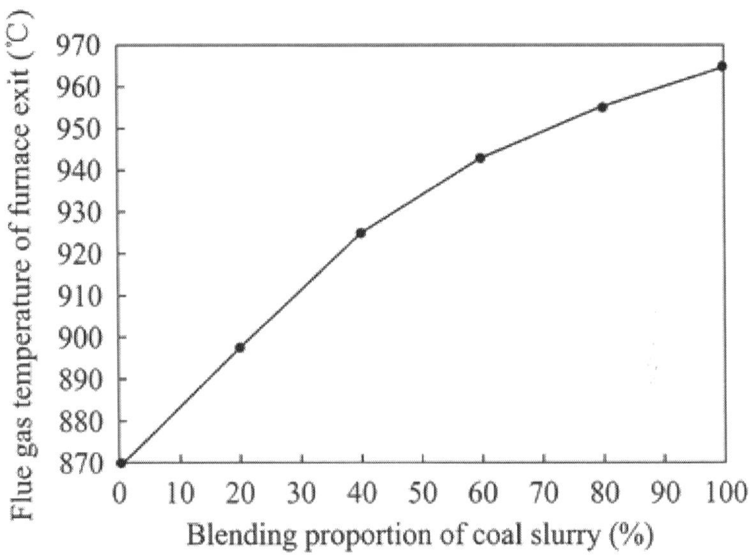

Figure 11. Relation between flue gas temperature of furnace exit and blending proportion of coal slurry

To meet a high blending proportion of slurry and improve the boiler thermal efficiency, a three-level superheater and a two-level desuperheater of the steam temperature regulated system are designed as follows.

The spray desuperheater is divided into two levels. The level 1 is installed at the entrance of the platen superheater. The level 2 is installed in the middle of the hot and cold section of the high-temperature superheater.

The level 1 desuperheater is designed as follows: Two level 1 desuperheaters with a water flow of 6.94 t/h, are used in series to increase the quantity of cooling water; and a new desuperheater is used to improve the efficiency of heat exchange. Combining these two methods can keep the inlet temperature of the platen superheater less than 400 °C.

3.7 Return fuel system

The return fuel system uses the steam cooled cyclone separator and a U return feeder as shown in Fig. 12. They are used to ensure the fuel to circulate and reduce the unburned fuel in the flue gas and increase the combustion efficiency. The main performance requirements of return fuel system are that the efficiency of the separator is less than 99 %, and the fluidization air pressure is not higher than 1.2 MPa.

Figure 12. The U return feeder

At the furnace exit we arrange two steam-cooled cyclone separators with Ø4400 mm in diameter. The cyclone separator takes tangential air inlet method to ensure the separation efficiency is no less than 99.5 %.

By analyzing all levels steam of the turbine, it can be discovered that the third level whose pressure is about 1.0 MPa and temperature is above 300 °C, simultaneously meets the requirements of pressure and temperature. Therefore, the third level is used to instead of compressed air as the fluidization dynamics of return fuel.

3.8 Ash conveying system

The system adopts a novel pneumatic pump device LFB150 to convey ash. Its operating parameters are shown in Table 5. The conveying pressure is between 0.04 MPa and 0.1 MPa, and the conveying pipe is Ø180 mm × 8 mm in diameter. Figure 13 shows the application of the LFB150.

Table 5. LFB pump process parameters

Name	LFB150
Conveying capacity	10–25 m^3/h
Air consumption	10–30 m^3/min
Airpressure requirements	0.03–0.1 MPa
Outlet pipe diameter	150 mm
Roots blower	FSR200:Q_s = 35.36 m^3/min
Conveying pipe	Ø194 mm × 6 mm, a total of 70 m

Figure 13. The application of the material sealing pump

4 COMBUSTION EFFICIENCY TEST

In order to investigate the overall performance of the boiler, and analyze its technical characteristics, experiments under different load conditions have been carried out. The criterions used are the standards of the People's Republic of China: (1) "Test code of thermal performance for Industrial boiler" (GB/T10180-2003), (2) "Monitoring and testing method for energy saving of industrial boiler" (GB/T15317-94), (3) "General specification for industrial boiler" (ZB/T10099-1999), (4) "The determination of particulates and sampling methods of gaseous pollutants from exhaust gas of stationary source" (GB/T16157-96) and (5) "Measurement method of smoke and dust emission from boiler"(GB5468-91), etc.

The two-stage ignition method is adopted during the boiler test process; when the boiler load reaches 40 %–50 % of the rated value, the temperature of smoke around the furnace outlet is 890–900 °C, the amount of slurry can be increase gradually. In accordance with the compress air parameters the throttle is adjusted to achieve the required load to ensure slurry stable combustion, then the startup process ends. During the boiler operating, the furnace pressure differential maintains between 500 Pa and 1000 Pa and the material layer pressure differential maintains between 7000 and 9000 Pa.

The ignition point of slurry is 1.66–1.87 times as high as that of the same kind of coal powder. The water evaporation absorbing heat makes the ignition time delayed significantly. The supply of compress air is the key factor to reduce the slurry ignition. The primary air with the 26 m/s speed adjusts the fluidization, furnace temperature and material layer pressure differential. When the load of the furnace reaches 50 t/h, the secondary air with a speed of 55 m/s is adopted to replenish oxygen. When the boiler load reaches the rated load, the ratio of primary air to secondary air is about 5:5.

The monitoring results of the boiler operating are shown in Fig. 14, the thermal efficiency parameters of the boiler are shown in Table 6, and the boiler emissions test results are shown in Table 7.

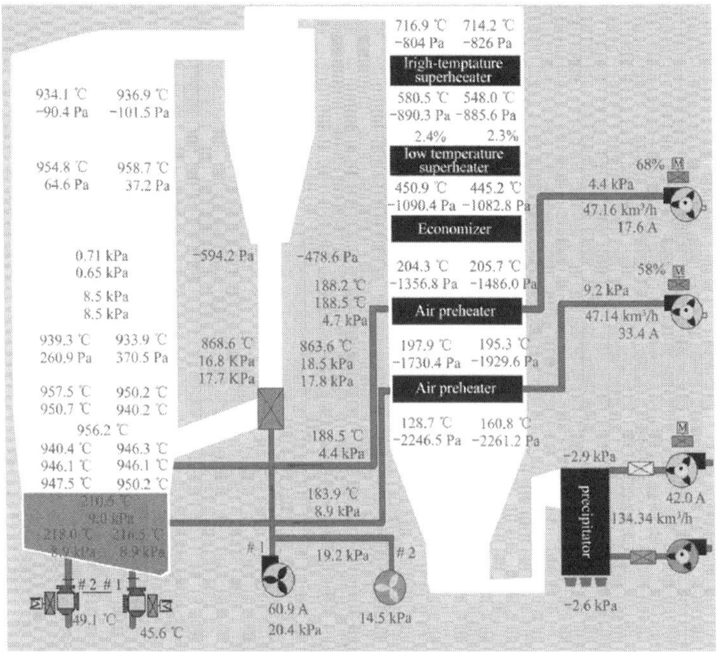

Figure 14. Operating parameters of slurry boiler

Table 6. Thermal efficiency parameters of the slurry boiler

Name	Unit	Rated test
Carbon content in fly ash	%	5.42
Carbon content in slag	%	2.75
O_2 in discharge flue	%	7.3
Discharge flue temperature	°C	145
Heat loss of discharge flue	%	9.11
Chemical incomplete combustion heat loss	%	0
Mechanical incomplete combustion heat loss	%	2.76
Dissipated heat loss	%	0.82
Total heat loss	%	10.17
Boiler thermal efficiency	%	89.83
Combustion efficiency	%	96.24

Table 7. Boiler emissions test

Name	Unit	Value
Slurry moisture	%	30
Counter balance efficiency of the rated load	%	88.46
Counter balance efficiency of 110 % rated load	%	87.24
Counter balance efficiency of 50 % rated load	%	86.63
Ash content of combustible (rated load)	%	5.42
NO_x emission concentration	mg/m^3	74
SO_2 emission concentration (no desulfurizer)	mg/m^3	35.9
Ringelmann level		0.5

From the testing results of the boiler, as shown in Fig. 14 and Table 6, it can be seen that all the operating parameters meet the requirements of design during boiler operation. The boiler has realized 89.83 % thermal efficiency, 96.24 % combustion efficiency, and 94 % the blending proportion of slurry.

As shown in Table 7, the emission of SO_2 is 35.9 mg/m^3,the emission of NO_x is 74 mg/m^3, and the Ringelmann level is 0.5. The pollutants emissions of the boiler are in accordance with "Emission standard of air pollutants for coal-burning oil-burning gas-fired boiler" (GB13271–2001).

The suspension-floating-circulating fluidized technology of boiler has been used in the Baizhuang mine of Shandong province of China; three CFB boilers are operating currently by using the suspension-floating-circulating fluidized technology. In 2012, the quantity of burned coal slurry was 462,300 t. Until now, the boiler has been used very successful in burning steadily and operating easily.

Experience has shown that the CFBB, with a capacity higher than 130 t/h, can also be retrofitted by using the suspension-floating-circulating fluidized technology, and the new proposed technology has a good development prospect.

5 CONCLUSIONS

A novel suspension-floating-circulating fluidized combustion technology was proposed, and applied in the 130/th CFBB. The retrofitted CFBB has solved some problems that exist in large-capacity CFBB for burning coal slurry. For example, MNS has achieved coal slurry transport that was fully enclosed, steady, long-distance and non-polluting, coal gangue as bed-material has reduced effectively the wear and tear of the boiler and improved the combustion efficiency of the boiler, the innovative sealed device has effectively reduced the pipe blockage and prolonged the life of the conveying belt, the multilevel asymmetry secondary air has provided adequate oxygen and ensured the combustion efficiency and desulfurization efficiency, two new type desuperheater before the platen superheater has realized a wide range of steam temperature regulation and keep the inlet temperature of the platen superheater less than 400 °C, and the steam as fluidizing air cannot only has avoided the secondary combustion of the returned fuel but also increased significantly the desulfurization efficiency, etc.

The experiments show that the new technology has realized 94 % the blending proportion of slurry,89.83 % thermal efficiency, and 96.24 % combustion efficiency. So the new technology provided an effective method to large-scale application of the slurry.

ACKNOWLEDGMENTS

The study is supported by the China-Africa international cooperation projects (2012DFG71060), the National Science and Technology Support Program in China (2012BAB13B04), South African National Research Foundation (78673).

REFERENCES

1. Bo Y, Huang ZY, Huang QX, Zhang YW, Zhou JH, Cen KF (2013) Combustion characteristics of coal-water slurry in a slag-tap vertical cyclone furnace through digital imaging. Energy Fuels 27(6):3427–3437

2. Cao XZ, Zhao WD, Liu JZ, Sun JF, Zhou JH, Cen KF (2009) A thermogravimetry investigation on the combustibility of coalslime water slurry. J China Coal Soc 34(10):1394–1399

3. Chen LY, Duan YF, Zhao CS, Yang LG (2009) Rheological behavior and wall slip of concentrated coal water slurry in pipe flows. Chem Eng Process 48(7):1241–1248

4. Cheng J, Zhou JH, Huang ZY, Liu JZ, Yang WJ, Cen KF (2008) Design and combustion of a 130 t/h coal washing sludge and water mixture-fired boiler with high steam temperature and pressure. J. Power Eng 28:367–376

5. Chi Y, Jiang XG, Ni MJ, Yang JL, Yan JH, Huang GQ, Li XD, Cen KF (2001) Study on co-firing process of coal-washery-sludge and coal-washery-rejects in fluidized beds. J China Coal Soc 26(2):204–208

6. Daviault SG, Ramadan OB, Matida EA, Hughes PM, Hughes R (2012) Atomization performance of petroleum coke and coal water slurries from a twin fluid atomizer, Daviault. Fuel 98:83–193

7. Gagarin SG, Gyul'maliev AM (2009) Theoretical principles of use of coal fractions with different densities for combustion. Solid Fuel Chem 43(1):17–24

8. Gunka V, Pyshyev S (2014) Lignite oxidative desulphurization. Int J Coal Sci Technol 1(1):62–69

9. Jakobs T, Djordjevic N, Fleck S, Mancini M, Weber R, Kolb T (2012) Gasification of high viscous slurry R&D on atomization and numerical simulation. Appl Energy 93:449–456

10. Ji DG, Wang ZN, (2007) Research on the suspension combustion rate of coal water slurry. Progress Min Sci Safety Technol Pts A and B: 1657–1661

11. Jiang MH, Xiao P (2009) Large circulating fluidized bed boiler technology. China Electric Power Press, Beijing

12. Li X, Xiao F (2005) The concept of using steam as substitution for HP fluid air in CFB boiler. Boiler Technol 36:10–12

13. Liu M, Duan YF (2009) Resistance properties of coal-water slurry flowing through local piping fittings. Exp Thermal Fluid Sci 33(5):828–837

14. Liu LY, Min FF, Zhang MX, Zhao Q (2012) Sliming characteristics of different density raw coal. J China Coal Soc 37(1):182–186

15. Lu P, Duan RL (2006) MNS coal slurry pipeline system design research. New Technol New Process 10:39–41

16. Ma XY, Duan YF, Liu M (2013) Atomization of petroleum-coke sludge slurry using effervescent atomizer. Exp Thermal Fluid Sci 46:131–138

17. Ni MJ, Li C, Fang MX, Wang QH, Luo ZY, Cen KF (2014) Research on coal staged conversion poly-generation system based on fluidized bed. Int J Coal Sci Technol 1(1):39–45

18. Qu ZB, Dai CH, Chen DB (2010) 440 t/h circulating fluidized bed boiler coal slurry feed equipment improvement and application. Energy Saving Environ Protect 7:38–40

19. Shu GP, Zhang YZ (2014) Research on the maceral characteristics of Shenhua coal and efficient and directional direct coal liquefaction technology. Int J. Coal Sci Technol 1(1):46–55

20. Tang XM, Yang JL, Jiang XG, Chi Y, Yan JH (2007) Rheological behavior in Xing long zhuang coal washery sludge pump system of CFB boiler. Journal of China Coal Society. 32(6):643–646

21. Wang H, Jiang XM, Zhang MX, Ma YF, Liu H, Wu SH (2010) A new fluidization-suspension combustion technologyfor coal water slurry. Chem Eng Process 49(10):1017–1024

22. Wang A, Song Q, Tu GM et al (2014) Influence of flue gas cleaning system on characteristics of PM2.5 emission from coal-fired power plants. Int J Coal Sci Technol 1:4–12

23. Wu M, Pan Y, Zhao GR, Hu WW, Yang Y (2009) Pressure distribution of the coal slurry in the pipeline transportation. J China Coal Soc 34(2):267–270

24. Yu HL, Zhang C, Liu JZ, Cen KF (2008) Experimental study of the atomizing performance of a new type of nozzle for coal water slurry. Energy Fuels 22(2):1170–1173

7

CHAPTER

Upper-Bound and Finite Element Analyses of Multi-Row Sprocket During Cold Semi-Precision Forging Process

Wangjun Cheng[1,2] , Chengzhong Chi[2] , Yongzhen Wang[2] , Peng Lin[2], Wei Liang[2]

[1] *China National Heavy Machinery Research Institute Co., Ltd., Xi'an 710032, China*

[2] *College of Materials Science and Engineering, Taiyuan University of Technology, Taiyuan 030024, China*

ABSTRACT

The cold semi-precision forging of a multi-row sprocket was investigated using upper-bound (UB) and finite element methods combined with experiments. Based on the design of a new tooth profile for the sprocket, a cold semi-precision forging process and a kinematically admissible velocity field for filling the die cavity were proposed. Using the UB method, the velocity fields of the sprocket billet in the forming process were divided theoretically and calculated. The process of forging a multi-row sprocket was simulated using the FEM package Deform-3D V6.1 to obtain the distributions of the velocity field and the effective stress field in filling the die cavity. Similar to the simulated results, the experiment on cold forging a 5052 aluminum alloy sprocket was successfully performed. By comparing the calculated (UB method), experimental and simulated load-stroke curves, the calculated and simulated results were basically in accordance with the experimental results.

The study provides a theoretical foundation for the development of the precision forging of multi-row sprockets.

Keywords: Multi-row sprocket Cold semi-precision forging Upper-bound method Finite element analysis

1 INTRODUCTION

A multi-row sprocket is a complicated part widely used in chain driving systems for mining machinery and equipment, such as scraper conveyers. The tooth profiles of multi-row sprockets are conventionally machined using a hobbing process, which wastes material and lowers their strength. Common failures of sprockets in the complex environment of a coal mine are excessive wears and tooth fractures (Wang et al. 2012). The chain and multi-row sprockets often bear larger static and dynamic loads in the impact-fluctuating work conditions of a scraper conveyer (Jiao et al. 2014). Therefore, improving the strength of multi-row sprocket plays an extremely important role in the driving system of mining equipment. Thipprakmas (2011) presented the fine-blanking process for manufacturing sprockets improving their strength with simple processing operations, reducing production time and costs. Takagi et al. (2009) used the technology of powder metallurgy to compact the housing sprocket while maintaining dimensional tolerances and strength. As a whole, the precision forging of multi-row sprockets is a complex process with severe plastic deformation (SPD) which improves the mechanical properties of forged parts.

In recent years, many researchers have reported the analysis process of forging gears using the upper-bound method and finite element method, but no attempt has been made to analyze the precision forging of multi-row sprockets. Upper-bound and FE methods are characterized by less computation time and reasonable accuracy (Can and Misirli 2008). Choi et al. (2000) proposed a mathematical method using the upper-bound method to forge a spur gear from a hollow billet. Choi and Choi (1998) also modeled a new tooth profile to obtain a more practical method to analyze the kinematically admissible velocity fields in external spur gear forging. Politis et al. (2014) established a simplified FE model to obtain the material flow behaviors and thickness distributions of forged Bi-metal gears. Compared with analyzing spur gear forging using the upper-bound and finite element methods above, the cold semi-precision forging with upper-bound and finite element methods for multi-row sprockets is a new approach to establishing similarity in the tooth profiles of the gear and sprocket.

This study aims to analyze the velocity fields of a multi-row sprocket billet based on the upper-bound method. The effective stress fields were investigated using the FEM software DEFORM-3D V6.1. The new forming processes were verified by experiments using aluminum alloy. The flow-ability of material, filling-quality of the dies and load-stroke curves were respectively obtained during the cold semi-precision forging of a multi-row sprocket. It is expected that this study will provide meaningful guidance in assessing the design of the die and determining the manufacturing process of multi-row sprockets.

2 COLD FORGING PROCESS

A multi-row sprocket possesses a complex structure, with three rows of tooth profiles on the circumference of cylindrical surface and two groups of flange grooves in the tooth profiles. A diagram of a multi-row sprocket part is shown in Fig. 1. The tooth profile is the main forming part in forging it. Currently, it is generally machined using a hobbing process to form the teeth. The cold precision forging processes of a multi-row sprocket are still in the experimental stage because of the complicated processing conditions. In this study, based on the modified sprocket forging profiles (Zhao et al. 2014), a new forging process is presented for the cold semi-precision forging of a multi-row sprocket. The order of processing is shown in Fig. 2.

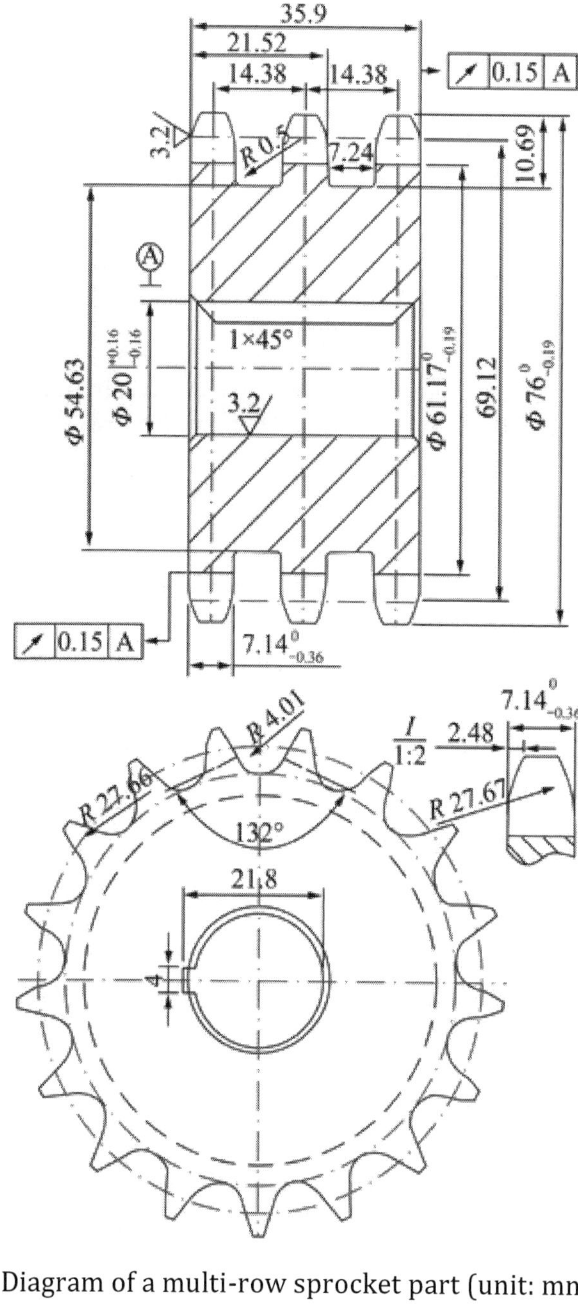

Figure 1. Diagram of a multi-row sprocket part (unit: mm)

Figure 2. Order of processing a multi-row sprocket

3 UPPER-BOUND ANALYSIS

3.1 Velocity fields assumption

A new simple type of sprocket tooth profile (Fig. 3) was designed to reduce the stress concentrations and cracks in the die cavity in cold semi-precision forging. The cylindrical coordinate (r,θ,z)(r,θ,z) system was used to analyze the velocity fields of the billet for the multi-row sprocket. The present theories were employed based on the upper-bound method with the following assumptions: (1) The friction coefficient is a constant value in the contact surface between the die and billet. (2) The material is perfectly rigid plastic, homogeneous and obeys Von Mise's yield criterion. (3) The billet volume remains constant in die cavity filling. (4) The billet filling is continuous without overlaps and cracks. (5) The external diameter of the hollow billet is equal to the root circle of the multi-sprocket.

3.2 Dividing deformation regions

There is a flow diversion surface with zero speed in forming the cylindrical hollow billet process, in which the material flows inwards inside the radius of the shunt circle and outwards outside the shunt circle (Ghassemali et al. 2013; Wu et al.2014). The outside diameter of the hollow billet is smaller than the root circle of the multi-sprocket. Therefore, the upsetting process in initially filling the die cavity was ignored to facilitate analysis (Wu and Wang 1992; Yang et al. 1999). The multi-row sprocket billet was divided into $2N$ deformation units and each unit $(N/2)$ was subdivided into five plastic deformation zones whose boundaries were ABCDEFGHI. It was assumed that billet fluids cannot cross or shear along the symmetrical planes AB and EFG. Hence half a tooth is used to analyze deformation characteristics throughout the cold semi-precision forging process.

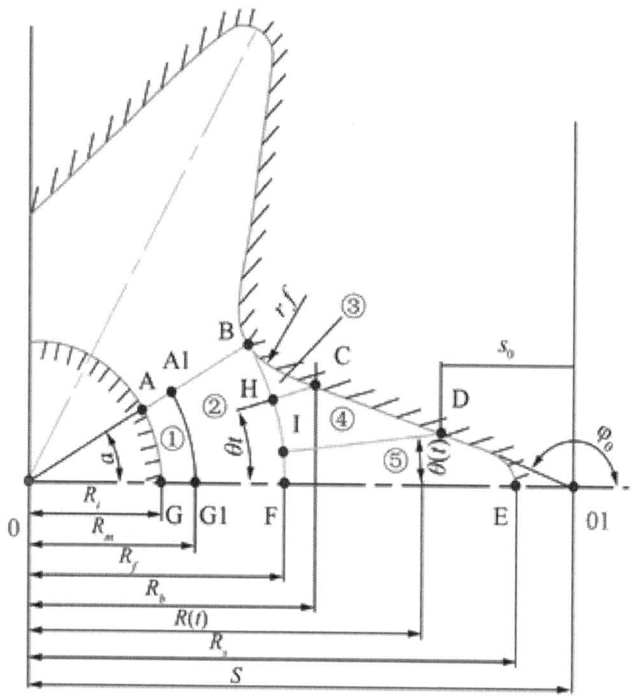

Figure3. Divided deformation regions for the cold forging of a sprocket

3.3 Kinematically admissible velocity

The velocity of the billet should satisfy the boundary conditions, the parallel velocity field assumptions and volume constancy, expressed as below:

$$\dot{\varepsilon}_{RR} = \frac{\partial U_R}{\partial R}, \quad \dot{\varepsilon}_{\theta\theta} = \frac{1}{R}\left(\frac{\partial U_\theta}{\partial \theta} + U_R\right), \quad \dot{\varepsilon}_{zz} = \frac{\partial U_z}{\partial z}, \tag{1}$$
$$\dot{\varepsilon}_v = \dot{\varepsilon}_{RR} + \dot{\varepsilon}_{\theta\theta} + \dot{\varepsilon}_{zz} = 0.$$

Assuming the upper die moves down with unit flow velocity u, the axial velocity $U_z$and the axial strain rate ε˙zzε˙zz were given as:

$$U_z = -\frac{u}{t}z, \quad \dot{\varepsilon}_{zz} = -\frac{u}{t}. \tag{2}$$

3.3.1 Deformation region ① (0≤θ≤α,Ri≤R≤Rm)(0≤θ≤α,Ri≤R≤Rm)

The radial velocity $U_R$was zero because of the constraint on the central mandrel surface AG and the excess metal flowed into region ② across the boundary A1G1. The circumferential velocity $U_\theta$was calculated assuming a constant volume. The velocity field and strain rates for this region were calculated as follows:

$$U_R = 0, \quad U_\theta = \frac{u\theta R}{t}, \quad \dot{\varepsilon}_{RR} = 0, \quad \dot{\varepsilon}_{\theta\theta} = -\frac{u}{t}. \tag{3}$$

3.3.2 Deformation region ② (0≤θ≤α,Rm≤R≤Rf)(0≤θ≤α,Rm≤R≤Rf)

The excess metal in region ① flowed into region ② across the boundary A1G1 and then into region ③④⑤ through the boundary BHIF. Therefore, boundaries A1G1 and BHIF satisfied the continuous conditions of deformation. The radial velocity $U_R$in the boundary A1G1 and the circumferential velocity $U_\theta$along the symmetric line were zero (Kwan 2002). The velocity distribution and the strain rates in this region were calculated as:

$$U_\theta = \frac{u\theta R}{t}\frac{a - \theta}{\alpha}, \tag{4}$$

$$U_R = \frac{u}{2t}\left(R - \frac{R_m^2}{R}\right). \tag{5}$$

Hence if θ=αθ=α, the calculated results were shown as follows:

$$U_\theta = 0, \quad \dot{\varepsilon}_{\theta\theta} = \frac{U_R}{R}, \tag{6}$$

$$U_R = \frac{u}{2t}\left(R - \frac{R_m^2}{R}\right), \tag{7}$$

$$\dot{\varepsilon}_{RR} = \frac{\partial U_R}{\partial R} = \frac{u}{2t}\left(1 + \frac{R_m^2}{R^2}\right). \tag{8}$$

3.3.3 Deformation region ③ (θf≤θ≤α,Rf≤R≤Rb)(θf≤θ≤α,Rf≤R≤Rb)

Assuming the fillet radius of the tooth groove circle for the sprocket is r_f, the velocity field in this region was assumed as follows:

$$U_R = \frac{uR}{2t} + \frac{c_1}{R}, \tag{9}$$

$$U_\theta = \left(\frac{uR}{2t} + \frac{c_1}{R}\right)\cot\varphi, \tag{10}$$

$$c_1 = \frac{uR_f^2}{2t}\left(\frac{\theta_f}{\alpha - \theta_f}\right) + \frac{uR_m^2}{2t}\left(\frac{\alpha}{\alpha - \theta_f}\right), \tag{11}$$

$$\dot{\varepsilon}_{RR} = \frac{\partial U_R}{\partial R} = \frac{u}{2t} - \frac{c_1}{R^2}, \tag{12}$$

$$\dot{\varepsilon}_{\theta\theta} = \frac{1}{R}\left(\frac{\partial U_\theta}{\partial\theta} + U_R\right) = \frac{u}{2t} + \frac{c_1}{R^2}, \tag{13}$$

where $\cot\varphi$ is a cotangent function associated with r_f, R_f, and R.

3.3.4 Deformation region ④ (θd≤θ≤θf,Rb≤R≤R(t))

To simplify the complicated relations, the translational coordinate system (Fig. 4) was constructed placing the point O_1 as a new coordinate origin. The tangential flow of metal in the deformation zone ④ was bound by molding of the inner surface CD, which satisfied the continuous condition between boundary CH and HI. If φ=φ0φ=φ0 or φ=πφ=π, the circumferential velocity U_θ was assumed to be zero. Thus the velocity and strain rate fields are given by:

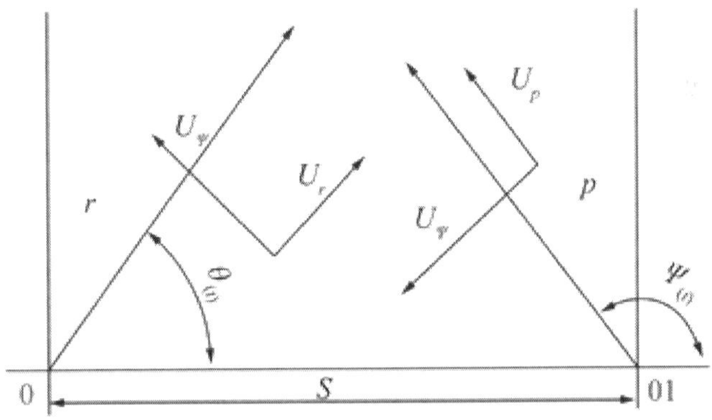

Figure4. Coordinates transformation diagram

$$U_\varphi = c_1\rho(\varphi - \varphi_0)(\varphi - \pi), \quad (14)$$

$$U_\rho = \left(\frac{\rho}{2} + \frac{c_3}{c_2\rho}\right)\left[\frac{u}{t} - c_1(2\varphi - \pi - \varphi_0)\right], \quad (15)$$

$$\dot{\varepsilon}_{\varphi\varphi} = \frac{1}{\rho}\left[c_1\rho(2\varphi - \pi - \varphi_0) + U_\rho\right], \quad (16)$$

$$\dot{\varepsilon}_{\rho\rho} = \frac{\rho^2 - 2}{2\rho^2}\left[\frac{u}{t} - c_1(2\varphi - \pi - \varphi_0)\right], \quad (17)$$

where c_1, c_2, c_3 represent the undetermined constants. The relationship between θ and φ can be characterized by the following equations:

$$r \sin\theta = \rho \sin(\pi - \varphi), \tag{18}$$

$$\rho^2 + s^2 - 2\rho s \cos(\pi - \varphi) = r^2. \tag{19}$$

The velocity condition of the boundary HI was obtained from:

$$U_\rho \cos(\theta - \varphi) + U_\varphi \sin(\theta - \varphi) = \frac{u}{2t}\left(r - \frac{r_m^2}{r}\right). \tag{20}$$

3.3.5 Deformation region ⑤ (0≤θ≤θd,Rf≤R≤Rs)

Because of the constraint of boundary DE, the radial velocity approximated to zero when billet fluids filled the die cavity. The excess metal in region ④ flowed into region ⑤ across the boundary ID, which satisfied the continuous condition on the boundary ID. It was observed that the width of the sprocket teeth was too small. Therefore, when p was equal to s_0, it was concluded that the radial velocity was approximately zero. The velocity field and strain rate respectively come out to be:

$$U_\varphi = (\varphi - \pi)\left[\rho\left(\frac{u}{t} + 2d_1\right) - d_1 s_0\right], \tag{21}$$

$$U_\rho = d_1(s_0 - \rho), \tag{22}$$

$$\dot{\varepsilon}_{\rho\rho} = -d_1 s_0, \tag{23}$$

$$\dot{\varepsilon}_{\varphi\varphi} = -\pi\left(\frac{u}{t} + 2d_1\right) + \frac{\pi d_1 s_0 + d_1(s_0 - \rho)}{\rho}. \tag{24}$$

3.4 Upper-bound solution

3.4.1 Power dissipation of deformation (E˙PE˙P)

$$\dot{E}_P = \int_V \bar{\sigma}\dot{\bar{\varepsilon}}\mathrm{d}V. \tag{25}$$

3.4.2 Power dissipation of friction (E˙fE˙f)

$$\dot{E}_f = \int_A \frac{m\bar{\sigma}}{\sqrt{3}}\Delta V\mathrm{d}A. \tag{26}$$

3.4.3 Power dissipation of shear (E˙SE˙S)

$$\dot{E}_S = \int_S \frac{\bar{\sigma}}{\sqrt{3}}\Delta V\mathrm{d}S. \tag{27}$$

The relative average forging load (P_{av}) is approximately given by:

$$\frac{P_{av}}{\bar{\sigma}} = \frac{\dot{E}_P + \dot{E}_f + \dot{E}_S}{AV_0\bar{\sigma}}, \tag{28}$$

Parameters used in these formulas are defined as the following:

t	Thickness of the billet;
z	Axial displacement of the punch;
d_1	An undetermined constant;
N	Number of teeth in the sprocket;
u	Unit flow velocity of the upper die;
$\dot{\varepsilon}_v$	Total strain rate of volume;
$R_i, R_m, R_f, R_b, R_s, s, s_0$	Distance from the sprocket center to the free surface of billet;
$R(t)$	Distance from the sprocket center to the free surface where the thickness of the billet is t;
$U_z, U_r, U_\theta, U_\rho, U_\varphi$	Velocities of axial, radial, circumferential directions, and in the direction of p and φ;
$\dot{\varepsilon}_{zz}, \dot{\varepsilon}_{RR}, \dot{\varepsilon}_{\theta\theta}, \dot{\varepsilon}_{\rho\rho}, \dot{\varepsilon}_{\varphi\varphi}$	Strain rate in the axial, radial, circumferential direction, and in the direction of p and φ;
r_f	Fillet radius of the tooth groove circle for the sprocket;
$\alpha, \theta_d, \theta_f, \varphi_0$	Angles in the radius;
$\theta(t), \varphi(t)$	Radius of free surface;
r, p	Radial distance from the coordinate origins;
A	Cross-section of the billet;
V_0	Upper die velocity;
m	Frication factor.

4 FINITE ELEMENT ANALYSIS

Based on the implicit Lagrangian FE code (Chandrasekhar and Singh 2011), the commercial FEM package DEFORM-3D V6.1 was applied to simulate the velocity field of cold semi-precision forging. Three-dimensional solid models were established using Pro/E 4.0 software, which were assembled and imported into the DEFORM-3D software in the form of STL files. Tetrahedral elements were used to mesh the grid number of 10,000 and the complete cold forging simulation was performed in 120 steps with displacements for the movement of the top die in each step of 0.15 mm. The velocities of the top die and floating bottom die were consistent at 15 mm/s by setting the processing of the velocity interface in the DEFORM-3D software. The 3D FE model of the cold semi-precision forging operation was constructed as shown in Fig. 5.

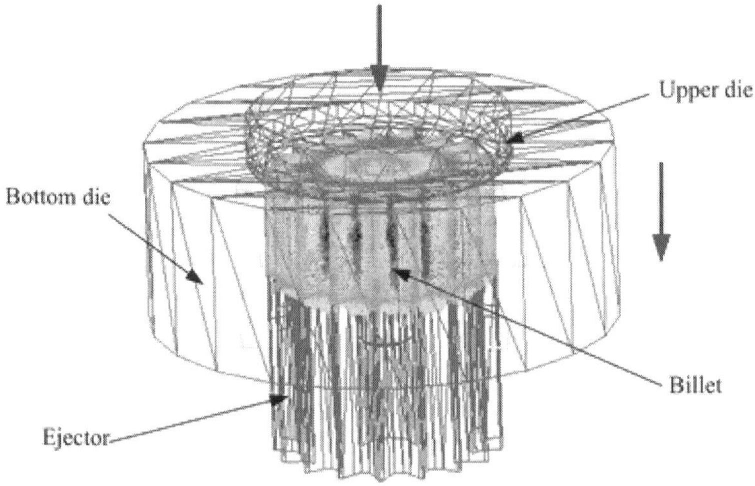

Figure 5. FEM of the multi-row sprocket in cold semi-precision forging

5 EXPERIMENTAL WORK

The cold semi-precision forging experiments for the multi-row sprocket were conducted using 5052 aluminum alloy. A constrained shunt of the core shaft for the hollow billet was applied to ensure the die cavity fully filled. The standard geometry sizes of the bottom die were cut using a wire EDM machine. The die components (die insert, upper die, bottom die and ejector) were produced from hot-work tool steel AISI H13. Generally, there were two alternative bottom dies to choose (the fixed bottom die and the floating bottom die). The main difference between these two bottom dies was in the effects of frictional forces in the interface between the billet and the bottom die (Dean 2000; Cai et al. 2004).

In design, the bottom die was fixed rigidly to the die holder, and did not move together with the upper die, which gave rise to a resistance from the frictional force. Conversely in the design of the floating bottom die, the upper and bottom dies moved together under the action of a spring, which caused the frictional force to be an effective forging force. The die structure for cold semi-precision forging of the multi-row sprocket was designed as shown in Fig. 6.

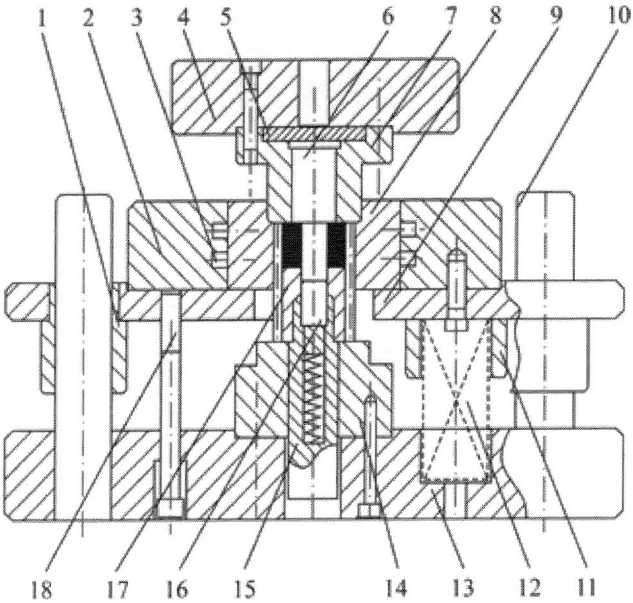

Figure. 6. Schematic of the die structure of a multi-row sprocket in cold semi-precision forging: *1* guide bushing; *2* loating die holder; *3* heating resistor; *4* upper platen; *5*support plate; *6* upper die; *7* upper die holder; *8* floating bottom die; *9* floating die holder; *10* guide pin; *11* spring retainer; *12* spring; *13* lower platen; *14* cushion block; *15* ejector beam; *16* locating pin; *17* ejector of sprocket; *18* hexagon socket head cap screws

A 500 ton hydraulic press was used to forge the sprocket at room temperature. Cylindrical hollow billets, whose initial geometry size was ϕ60×ϕ18×52ϕ60×ϕ18×52 and initial test temperature was 20 °C, were lubricated with oil-base graphite. The velocity controller of the hydraulic press was set to fix the forming velocity of the hydraulic press at 15 mm/s. The displacement of the upper die was measured with a displacement sensor and the load on the hydraulic press was obtained with a load sensor.

6 RESULTS AND DISCUSSION

6.1 Simulation results

Based on the analysis of the upper-bound method, the velocity field for the cold semi-precision forging process for a multi-row sprocket was verified by FE simulation to obtain the distributions of the velocity fields in the die cavity filling. Figure 7 indicates the distribution of the velocity field of the hollow billet. The maximum velocity in region ⑤ was 321 mm/s and the minimum velocity in region ① was close to zero (Fig. 8). The fluid particles of the billet were given priority to flow radially and convex ears of the velocity outside the hollow billet occurred when filling the die cavity (Fig. 7b, c, d). The flow velocity of the billet increased sharply in the tooth-groove of the die cavity because of the path with least resistance when the billet made first contact with the die cavity. Because the velocity increased exponentially, the billet fluids fully filled the die cavity. The self-locking of the billet in filling the die cavity resulted in a sharp decrease in the velocity. At the end of the forging stage, irregular fluctuation of the billet fluids occurred in the teeth of the sprocket, as observed in Fig. 7d.

The distribution of the effective stress field reflected the filling resistance and stress concentrations of the deformation billet (Bennett 2013; Wang et al. 2011). Figure 9 shows the distribution of the effective stress field at various steps. As the upper die travel increased, it was obvious that the billet compressed axially and bulged radially. The maximum stress of 213 MPa was found in the teeth of the sprocket billet and higher stress concentrations were indicated at the edges and in the shunt hollows of the billets. A minimum stress of 26.3 MPa was observed in the middle of the sprocket billet. At the end of the forging stage, the restraint effect of

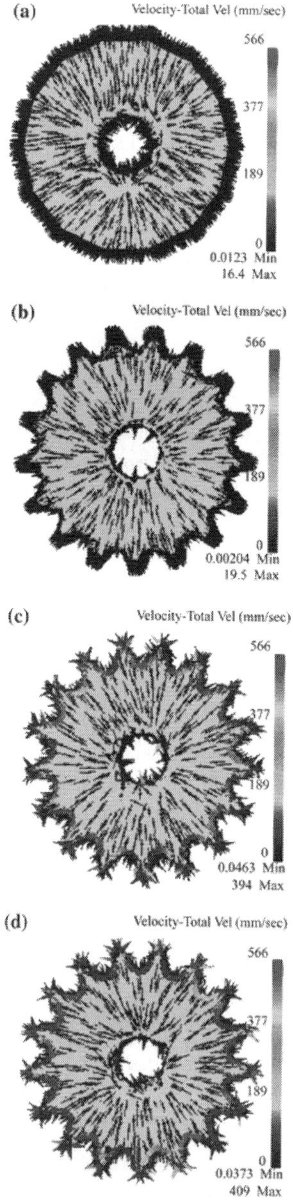

Figure 7. Distribution of the velocity field at different steps during the cold forging process**a** 34th step; **b** 64th step; **c** 84th step; **d** 104th step

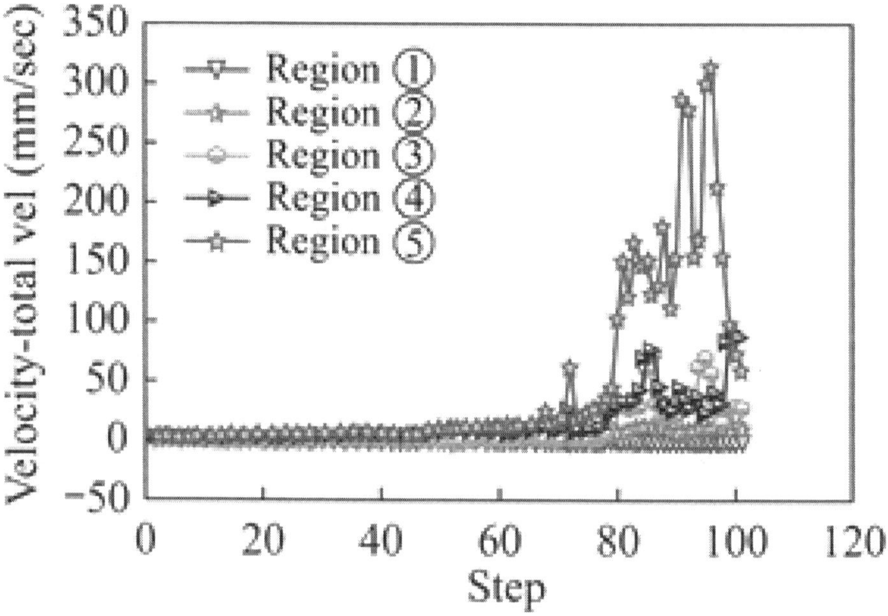

Fiure 8. Variation in velocity at various steps in various deformation regions

the dies reached a maximum, which made the billet flow more difficultly. Larger rough selvedges occurred at the bottom of the billet corner. Figure 10 illustrates the variation in effective stress at various steps. It was observed that the effective stress increased exponentially in the initial stage and then remained constant at the end of forging. The value of effective stress in region ⑤ remained the highest at 213 MPa, while region ① remained the smallest. The highest effective stress values were recorded for the teeth of the sprocket because of the severe plastic deformation (SPD). The effective stress of the sprocket decreased from the teeth to the middle of the billet and increased from the middle to the shunt hollows of the sprocket billet.

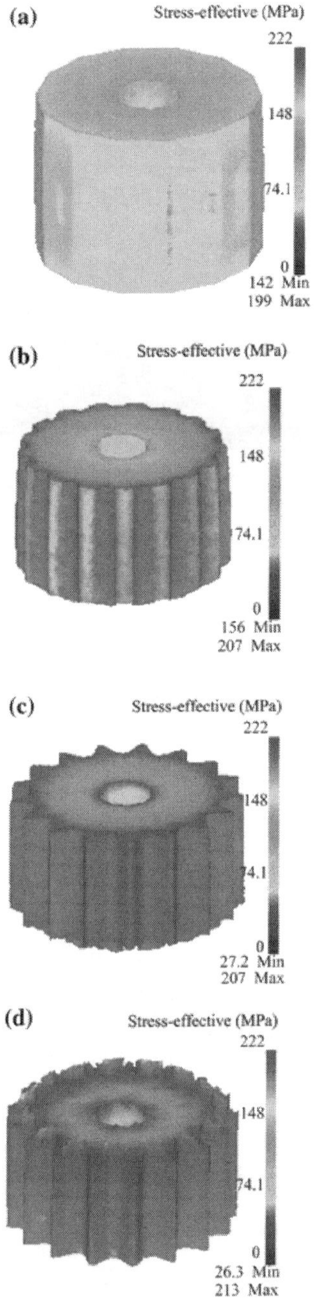

Figure 9. Effective stress distribution with various steps **a** 34th step; **b** 64th step; **c** 84th step; **d** 104th step

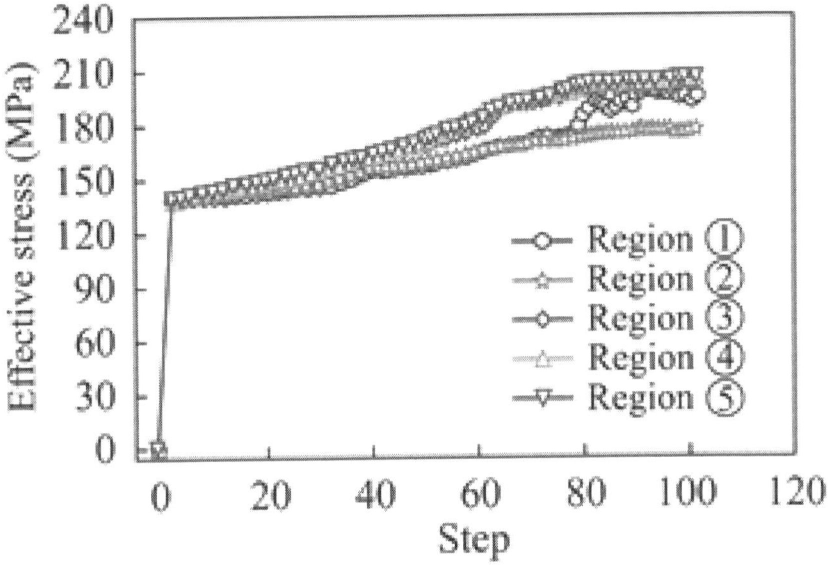

Figure 10. Variation in effective stress with various steps

6.2 Experimental results

The simulation results for the forging process provided significant guidance to the practical experiments. It was found that the tooth profile of the forged sprocket in the experiment (Fig. 11) was fully filled and the geometries of a forged sprocket with the new process possessed good integrity. The similarity in the tooth profile between the experimental and simulated results was obvious. It showed that the results with the new forging process completely fulfilled the engineering requirements in precision forging a multi-row sprocket.

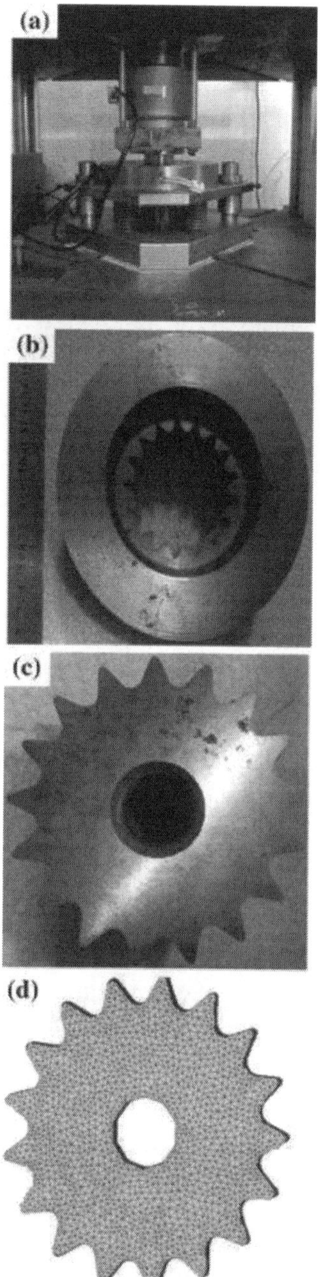

Figure 11. Photos of the experimental dies and forged sprockets **a** die structure with load sensor; **b** *bottom* die; **c** forged part; **d** simulated part

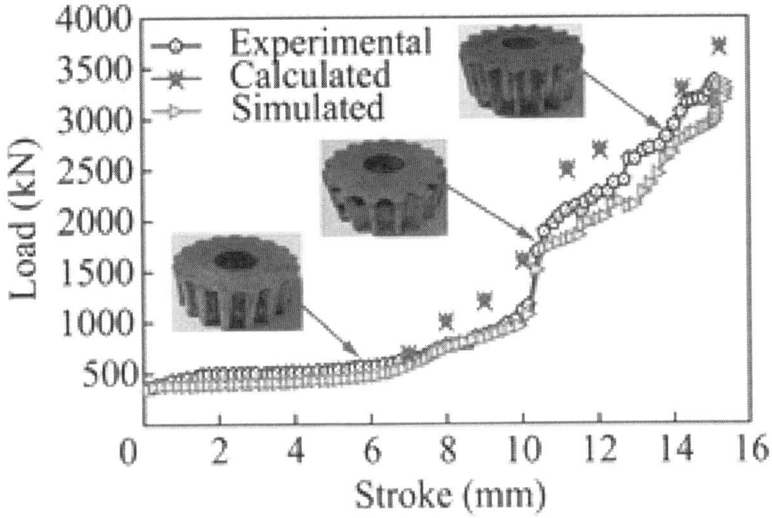

Fig. 12. Load-stroke curves during cold forging of multi-row sprocket

6.3 Forging loads

It was inferred that the velocity discontinuity of the billet was an important factor to take into account in terms of the required forging load. This was extremely significant for designing accurate dies, choosing reasonable equipment and determining a proper process (Can and Misirli 2008; Liu and Cui 2009). The calculated, experimental and simulated load-stroke curves during cold semi-precision forging are illustrated in Fig. 12. It was observed that their variation trends were basically the same and the forging loads increased abruptly owing to the forming of sprocket teeth. The maximum forging load for the experiment was 3399.9 kN, which was less than the equipment load capacity of 5000 kN, while the maximum load calculated with the upper-bound method was 3700 kN and simulated was 3335.8 kN. The relative average forging load obtained using the upper-bound method was greater than from the experimental and simulated results because of the complex boundary conditions, shear loss, friction resistance and deformation energy, etc. The experimental loads were greater than the simulated loads because of the complicated experimental conditions, such as temperature control, friction effect, lubrication, and equipment, which the simulated results did not fully consider. The whole forging process was divided into four stages: free upsetting, filling, corner filling and recovery. The load was relatively lower in the free upsetting stage. In the filling and corner filling stages, the load increased abruptly with the increase in the stroke movement. Finally, the load tended to be stable in the recovery stage. The experimental results were in accordance with the calculated and simulated results, providing a level of validation for the accuracy of the theoretical results.

7 CONCLUSIONS

1. The cold semi-precision forging process for a multi-row sprocket and a kinematically admissible velocity field for filling the die cavity were presented and verified to be simple and feasible in practical application.

2. The validation of the upper-bound method was undertaken by comparing its results with FEM results. The velocity and effective stress fields were obtained and indicated that the upper-bound and FE methods were suitable for analyzing complex three-dimensional plastic deformation.

3. The cold forging experiments of the multi-row sprocket were successfully carried out at room temperature. The experimental load-stroke curves were basically in accordance with the calculated and simulated results. This study provides a theoretical foundation for research on the precision forging of a multi-row sprocket.

ACKNOWLEDGMENTS

This research was financially supported by the National Science Foundation of China (Grant Nos. 51175363 and 51274149).

REFERENCES

1. Bennett CJ (2013) A comparison of material models for the numerical simulation of spike-forging of a CrMoV alloy steel. Comput Mater Sci 70:114–122
2. Cai J, Dean TA, Hu ZM (2004) Alternative die designs in net-shape forging of gears. J Mater Process Technol 150:48–55
3. Can Y, Misirli C (2008) Analysis of spur gear forms with tapered tooth profile. Mater Des 29:829–838
4. Chandrasekhar P, Singh S (2011) Investigation of dynamic effects during cold upset-forging of sintered aluminium truncated conical preforms. J Mater Process Technol 211:1285–1295
5. Choi JC, Choi Y (1998) A study on the forging of external spur gears: upper-bound analyses and experiments. Int J Mach Tool Manuf 38:1193–1208
6. Choi J, Cho HY, Jo CY (2000) An upper-bound analysis for the forging of spur gears. J Mater Process Technol 104:67–73
7. Dean TA (2000) The net-shape forming of gears. Mater Des 21:271–278
8. Ghassemali E, Tan MJ, Jarfors AEW, Lim SCV (2013) Optimization of axisymmetric open-die micro-forging/extrusion processes: an upper bound approach. Int J Mech Sci 71:58–67
9. Jiao HZ, Yang ZJ, Wang SP (2014) Contact dynamics simulation analysis for sprocket transmission system of scraper conveyor. J Chin Coal Soc 39(1):166–171
10. Kwan CT (2002) An analysis of the closed-die forging of a general non-axisymmetric shape by the upper-bound elemental technique. J Mater Process Technol 123:197–202
11. Liu J, Cui ZS (2009) Hot forging process design and parameters determination of magnesium alloy AZ31B spur bevel gear. J Mater Process Technol 209:5871–5880

12. Politis DJ, Lin J, Dean TA, Balint DS (2014) An investigation into the forging of Bi-metal gears. J Mater Process Technol 214:2248–2260

13. Takagi M, Suganaga K, Nagata T (2009) New PM sprocket meets auto cost, performance concerns. Met Powd Rep 64:25–29

14. Thipprakmas S (2011) Improving wear resistance of sprocket parts using a fine-blanking process. Wear 271:2396–2401

15. Wang JH, Chen FX, Deng XZ (2011) Numerical simulation of hypoid driving gear precision forging. Forg Stam Technol 36:149–154

16. Wang SP, Yang ZJ, Wang XW (2012) Wear of driving sprocket for scraper convoy and mechanical behaviors at meshing progress. J Chin Coal Soc 37(2):494–498

17. Wu ZM, Wang HY (1992) An upper-bound analysis of shunt forming of gears. Corp Plast Pro Tech 23:306–310

18. Wu YJ, Dong XH, Yu Q (2014) An upper bound solution of axial metal flow in cold radial forging process of rods. Int J Mech Sci 85:120–129

19. Yang SH, Fu PF, Huang LJ (1999) Upper-bound model establishment and analysis of cold precision forging of spur gears. J Jilin Univ Technol 29:23–28

20. Zhao RH, Chi CZ, Wei CJ, Cheng WJ, Liang W (2014) Semi-precision hot forging blocking and die design of multi-row chain wheel. Forg Stam Technol 36(6):20–23

8

CHAPTER

A Study of Particles Penetration in Sieving Process on a Linear Vibration Screen

Zhanfu Li[1], Xin Tong[1]

[1] *School of Mechanical Engineering and Automation, Huaqiao University, Xiamen 361021, China*

ABSTRACT

This paper presented an investigation of particle collision and penetration using the discrete element method to understand the motion of particles and improve theoretical treatment in the sieving process. The process progressively was divided into looseness, stratification, collision, and penetration. Particle penetration has a direct effect on the screening performance. The penetration probability was defined, and the mathematical relationships between particle penetration and vibration parameters were established using the least squares method. To obtain the ideal penetration probability for materials the amplitude and frequency should preferably be near 3.0 mm and 25 Hz, respectively. The vibration direction angle has only a slight effect on penetration. The stage of the screening process from 0.1 to 0.7 s is the primary region for collision and penetration. This paper focused on the sieving process to more fully understand how particle collision and penetration influence the screening efficiency.

Keywords: Penetration probability Collision Discrete element method Sieving process

1 INTRODUCTION

The understanding of screening process is incomplete regarding the intricacies of particle-particle and particle-machine interactions. But the influence of these interactions on the screening process effects on both the efficiency and the value of the product (Soldinger 2000; Zhao et al. 2000; Jiao et al. 2006). We divide the screening process into four parts as follows: (1) looseness: the void space that occurs between particles which are rebounded by the sieve plate. A sufficient void space contributes to particle stratification (Zhang and Tong 2012). (2) Stratification: the ability of the fine particles to penetrate the coarse granules, when all particles are scattered. In other words, stratification relates to the ability of the fine particles to approach the screening surface (Xiao and Tong 2012). (3) Collision: once close to the screening surface, some particles impact the sieve wires and some pass through the sieve mesh. These particles that are near the mesh are referred to as the collision particles (Wei et al. 1999). (4) Penetration: the penetration rate is the fraction of the collision particles which penetrate the mesh. It has a direct effect on screening performance.

Some previous research has developed our current understanding and theory of single grain screen-penetrating probability. The probability model and equations for a spherical particle penetrating a hole were developed by Subasinghe et al. (1989). Wen and Liu (1982) put forward various characteristic motions that a single particle on the sieve surface may undertake: relative static, relative sliding and projectile motion. Applying chaotic motion theory to analyze the movement of the single particle, Liu and Lu (2009) concluded that the motion is nonlinear. Beyond the single particle, theory of particle cluster screen-penetrating probability has also been established. In 1972, the collision speed transfer formula of particle clusters in the sieving process was established by Bruder. Using the method of probability statistics, the mathematical model of the penetration probability along the screen was put forth by Chen and Yan (1982). In recent years, discrete element method (DEM) allows for modeling at the particle level and accounts for interaction between particles and boundary objects. Cleary (2004) and Cleary and Sawley (2002) presented a three dimensional model of spherical particle separation on an inclined flat screen based on DEM. Tong explored the relationship between vibration parameters and screen efficiency, as well as performing simulations of stratification and penetration over various structural and vibration parameters based on the DEM (Chen and Tong 2009, 2010; Wang and Tong 2011). The implementation of DEM for screening process gives an insight to study the particulate systems. DEM simulations have been reported positive results in mineral screening and crop cleaning (Zhu et al. 2008; Weerasekara et al. 2013; Cleary and Dsinnott 2015; Cleary 2015; Xue et al. 2015).

Taking into account particle-particle and particle–machine interactions, we study herein a series of screening simulations which provides a theoretical foundation for apparatus design based on optimized screening parameters, and provides a reasonable explanation for the screening process.

2 SIMULATION

Successful application of the DEM, which is a numerical method suitable for tracking particle characteristics including speed and displacement, to mineral processing and material separation represents a good way to optimize design of vibration screens. However, few studies about DEM simulations of the screening process have been reported.

2.1 The contact model used in the simulation

The model is based on the work of Mindlin (Mindlin and Deresiewicz 1953). The elastic contact occurs in two spherical granules whose radii are R_1 and R_2 respectively. R^* is the equivalent radius. The normal overlap, α is given by

$$\alpha = R_1 + R_2 - |r_1 - r_2| \tag{1}$$

$$\frac{1}{R^*} = \frac{1}{R_1} + \frac{1}{R_2} \tag{2}$$

with the centers of the two particles position vectors are given by r_1 and r_2. The normal force F_n, is given by

$$F_n = \frac{4}{3} E^* \sqrt{\alpha^{\frac{3}{2}} R^*} \tag{3}$$

$$E^* = \frac{1-\gamma_1^2}{E_1} + \frac{1-\gamma_2^2}{E_2} \tag{4}$$

where E^* is the equivalent Young's modulus. E_1, γ_1 and E_2, γ_2 are the Young's modulus and Poisson's ratio for particles 1 and 2, respectively. Additionally there is a damping force, FdnFnd, given by

$$F_n^d = -2\sqrt{\frac{5}{6}} \beta V_n^{\text{rel}} \sqrt{S_n m^*} \tag{5}$$

$$\beta = \frac{\ln e}{\sqrt{\ln^2 e + \pi^2}} \tag{6}$$

$$S_n = 2E^* \sqrt{\alpha R^*} \tag{7}$$

$$V_n^{\text{rel}} = (v_1 - v_2) n \tag{8}$$

where m^* is the reduced mass, VrelnVnrel is the normal component of the relative velocity. The two particles' speeds are v_1 and v_2. The tangential force, F_t depends on the tangential overlap δδ, and the tangential stiffness, S_t.

$$F_t = -\delta S_t \tag{9}$$

$$S_t = 8\left(\frac{2-\gamma_1^2}{G_1} + \frac{2-\gamma_2^2}{G_2}\right)\sqrt{\alpha R^*} \tag{10}$$

additionally, tangential damping is given by:

$$F_t = -2\sqrt{\frac{5}{6}}\beta V_t^{\mathrm{rel}}\sqrt{m^* S_t} \tag{11}$$

where VreltVtrel is the relative tangential velocity. G_1 and G_2 are the shear modulus of particles 1 and 2, respectively.

2.2 Simulation model and conditions

The 3D vibration screening model was set up to simulate the linear vibration based on DEM. In simulations, the particles were produced by particle factory (Fig. 1). The mode of screen surface was set by software EDEM. The discharge port saved the particles which did not pass the aperture. This paper focused on particles collision and penetration over the screen surface in the process of screening (Fig. 2). The three-dimensional (3D) model was used to simulate the movement of all particles in the screening process (Fig. 3). Because interactions happen between the particle and the screen surface, the spring-dashpot-slider model was used to simulate particles collision (Soliding 1999). The DEM model has been shown to provide good agreement between numerical and experimental results (Dong et al.2009). The screen body is 162 mm long, 32 mm wide and 80 mm high. Meanwhile, considering the size distribution of sands used in the industry, this paper employed a bimodal normal distribution of particles which are provided by the particle supply (Fig. 1) with a standard deviation of 0.545. The particles are spherical sand with mean diameters of 0.5 and 1.0 mm. As shown in Fig. 2, ββ is the vibration direction angle; αα is the inclination of the screen surface; *A* is the amplitude. In Fig. 1 the position of the particles are presented. In this model, the particles fall down to sieve plate constantly by gravity. The process of sieving simulation was presented on Fig. 3. The vibration parameters are mainly frequency, amplitude and vibration direction angle. Simulation conditions are summarized in Tables 1 and 2. Three type experiments have been accomplished. The particle generate rate is 42,000 particle/s, the screen declination is 21°, screen vibration is sinusoidal translation, the total number is 30,000. Each groups of experiments only changed one parameter value and the conditions are listed in Table 3. Such as group one, only amplitude is varying.

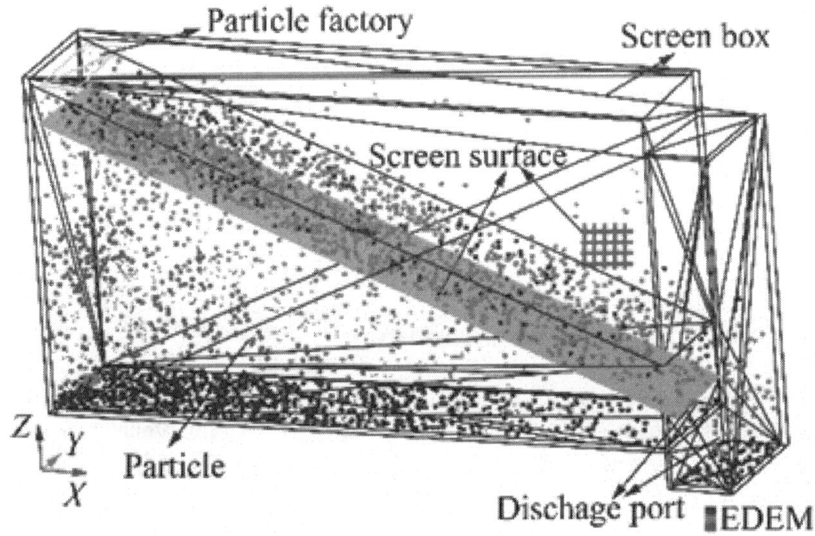

Figure 1. DEM-3D model

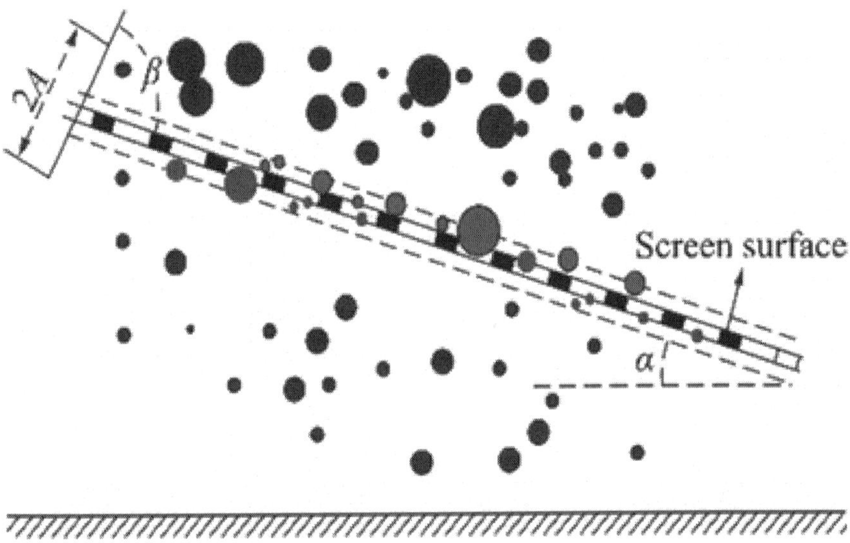

Figure 2. Schematic side-view of the screen surface

Figure 3. Process of sieving simulation

Table 1. Material properties setting in EDEM

Material properties	Poisson's ratio	Shear modulus	Density (kg/m^3)
Particles	0.3	23 MPa	2678
Screen	0.29	79.92 GPa	7861

Table 2. Collision properties setting in EDEM

Collision properties	Coefficient of restitutions	Coefficient of static friction	Coefficient of rolling friction
Particle-particle	0.1	0.545	0.01
Particle-screen	0.2	0.5	0.01

Table 3. Summary of conditions

Text group	Amplitude (mm)	Frequency (Hz)	Vibration angle
Group one	0.5, 1.5, 2, 2.55, 3.5	20	100
Group two	2.55	10, 20, 25, 30, 40, 50	100
Group three	2.55	20	10, 20, 30, 40, 50, 60, 70, 80

3 DEFINITION OF PENETRATION PROBABILITY AND DISCUSSION

During the screening process simulation, the particles took 0.3 s to move to the end of the screen surface and took over 1.3 s to complete screening. In this study, the penetration probability is defined as the ratio of the mass of those particles which penetrate the sieve plate to the total mass of all collision particles over the time period from 0.4 to 0.6 s. From the Fig. 2, the blue particles are called the penetration particles and the particles between the two *dotted lines* are the collision particles. The distance between the *dotted line* and the screening surface is the diameter of the largest particle. Because the penetration and collision particles have a direct effect on efficiency, the penetration and collision particles vary in different vibration parameters. The penetration probability is given by

$$T = \frac{M_\mathrm{p}}{M_\mathrm{c}} \times 100\ \% \tag{12}$$

where M_p is the penetration particles' mass, M_c is the mass of the collision particles.

This study forces on the effects of vibration parameters, including amplitude, frequency and vibration direction angle, on the penetration probability.

3.1 Frequency effects on penetration probability

Figure 4a shows the number of collision particles over time at different frequencies. At high frequency ($f > 40$ Hz) the changes in the particle number is not as obvious as at the lower frequencies and few particles have an opportunity to penetrate the apertures at high frequency (Fig. 4b). The time from the beginning of charging to 0.7 s is the main period of collision and penetration. The relationship between frequency and penetration probability is presented in Fig. 4c. The penetration probability decreases as the size of separated particles increases and the maximum probability occurs at a frequency between 20 and 25 Hz.

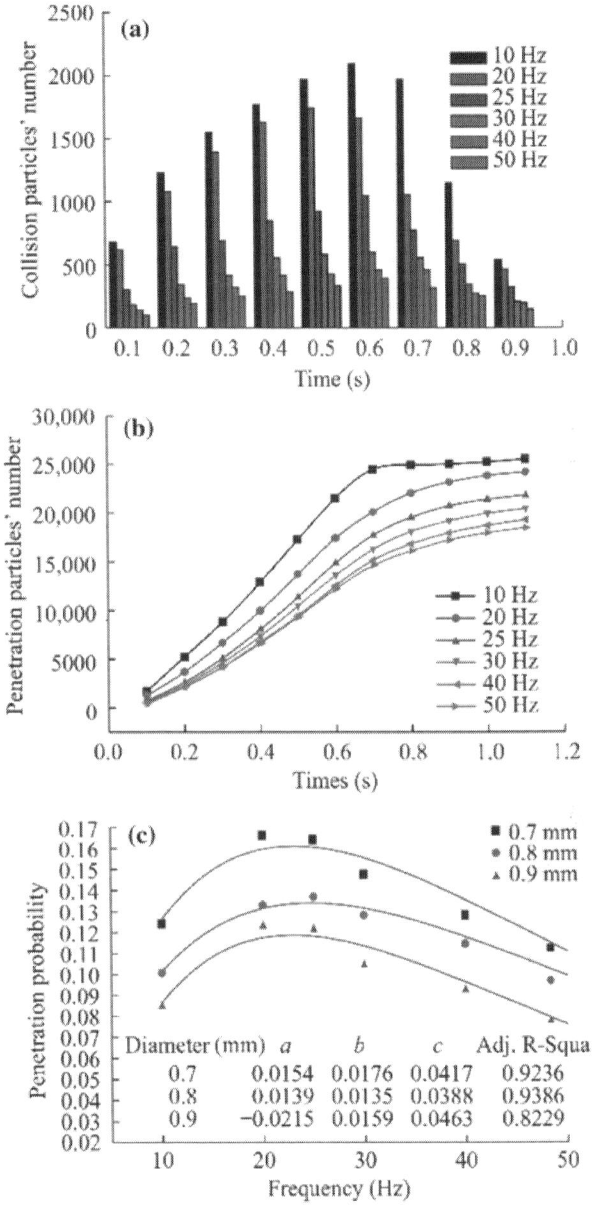

Figure 4. Effect of vibration frequency on penetration. **a** Number of collision particles over time, **b** Number of penetration particles over time, **c** Penetration probability versus frequency

From Fig. 4b, c, it can be seen that the mass of penetration particles increases, but the penetration probability decreases when the frequency is greater than about 25 Hz. This illustrates that the increment speed of the collision particles is greater than the penetration particles. Meanwhile, because the increment speed of the collision particles is less than that of the penetration particles when the frequency is less than 20 Hz, the penetration probability curves have an upward trend. When $f > 40$ Hz, the particles completely fill the screening box. At last, the data in Fig. 4c were fit, with low relative error, to the function:

$$y = (a + bx)e^{-cx} \tag{13}$$

where y is penetration probability, x is frequency, a, b, and c are fitting parameters, depending on particle separation diameter (Fig. 4c).

3.2 Amplitude effects on penetration probability

Figure 5a shows the number of collision particles over time for different amplitudes. At low amplitudes, because most particles are close to the sieve plate, the collision numbers obviously increase. At high amplitudes, the particles can gain more energy to be rebounded. The contacts between the sieve plate and particles decrease. The number of penetration particles over time is shown in Fig. 5b. The time from the beginning of charging to 0.7 s is the main region of collision and penetration, after which the slope of the change in number of penetration particles decreases. As the amplitude increases, the number of penetration particles decreases. We chose the statistical window to determine the penetration probability to be from the time at which the most particles are scattered evenly to the end time of feeding, that is, from 0.4 to 0.6 s.

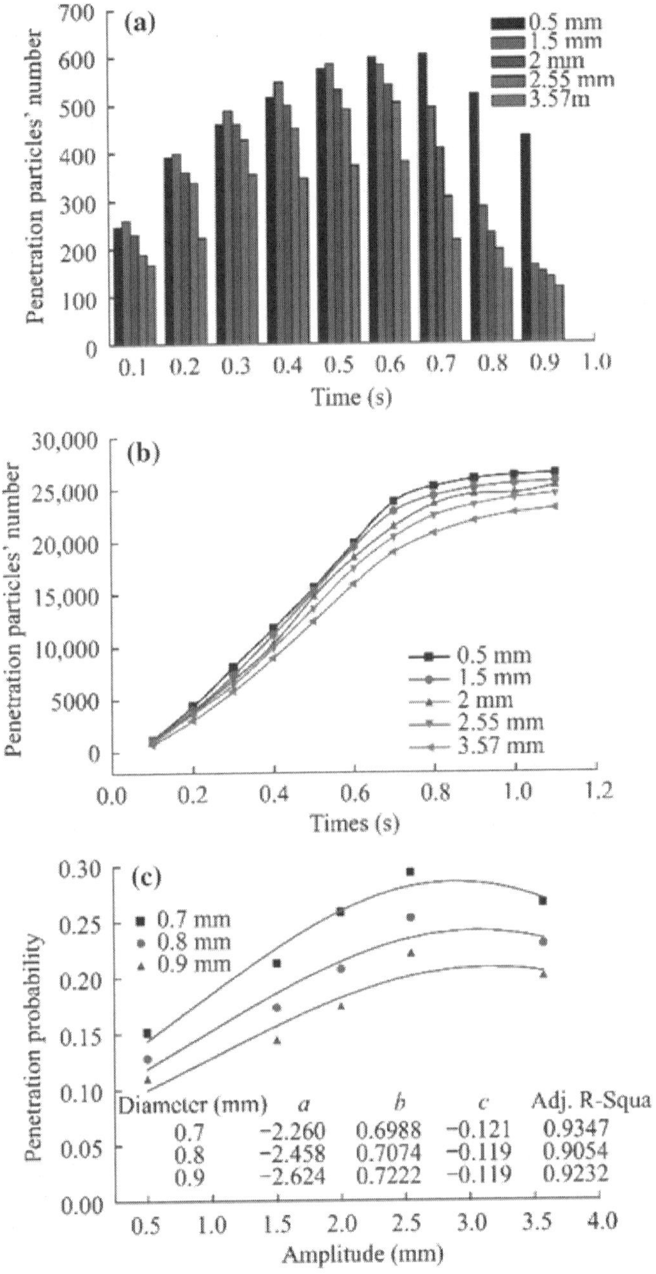

Fig. 5 Effect of vibration amplitude on penetration. **a** Number of collision particles over time, **b** Number of penetration particles over time, **c** Penetration probability versus amplitude

From Fig. 5c we see that the amplitude of 2.55 mm gives the highest value. Because of the speed of change about the numerator and denominator of Eq. (12), the curve presents that as shown in Fig. 5c. Data obtained with amplitudes less than 3.55 mm were fit to the following function, with low relative error:

$$y = e^{cx^2+bx+a} \tag{14}$$

where x is the amplitude, y is the penetration probability, and a, b, and c are fitting constants dependent on the particle diameter (Fig. 5c).

3.3 Vibration direction angle effects on penetration probability

Figure 6a demonstrates that the mass of the collision particles does not change much for different vibration direction angles. Unlike amplitude and frequency, the particles do not gain enough energy to jump and be thrown over the screen surface, and the vibration direction angle has only a slight effect on the mass of the collision. Figure 6a shows at the angle of 20 and 31, the particles' velocity component, which is perpendicular to screen surface is small, so they return to the sieve plate quickly. The change of the penetration particles' number over time is presented on Fig. 6b. In addition the difference is little among these different vibration angles. On the whole, the effect of vibration angle differs from that of the amplitude and frequency. The vibration angle has a more subtle effect than others (Fig. 6c).

The relationships between penetration probability and vibration parameters were presented. The main conclusions show that to obtain the ideal penetration probability for materials the amplitude and frequency should be chosen to be about 3.0 mm and 25 Hz, respectively. The vibration direction angle has a slight effect on the penetration probability. This paper focuses on the particles penetration under various single parameter experiments using the DEM and provides insight about screening process and a theoretical foundation for designing the screen equipment.

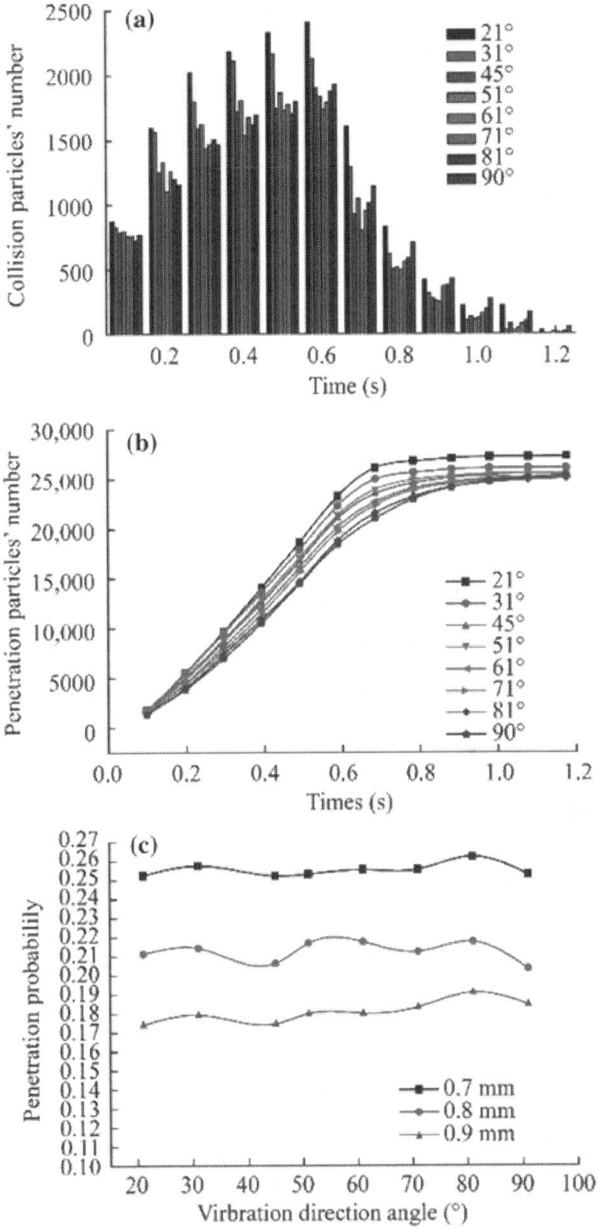

Figure 6. Effect of vibration direction angle on penetration. **a** Number of collision particles over time, **b** Number of penetration over time, **c** Penetration probability versus vibration direction angle

4 CONCLUSIONS

1) Simulations were used to establish changes in the number of penetration and collision particles over time. Because of the complex sieving process which is hard to gain the information from the sieving experiments, we simulated these experiments using the method of DEM which has been shown to provide good results for similar problems. The effects of vibration conditions on the sieving processes have been studied.

2) The results show that the stage of the screening process from 0.1 to 0.7 s is the main part for collision and penetration. The number of penetration particles mounted and leveled off. In a representative continuous screening process, the relationships between the penetration probability and the vibration parameters were modeled. To obtain the ideal penetration probability for materials the amplitude and frequency should be chosen to be about 3.0 mm and 25 Hz, respectively. The vibration direction angle has a slight effect on the penetration probability.

3) The sieving process was divided into four stages as follows: looseness, stratification, collision and penetration. Each part is important to screen performance. Further research will focus on the sieving process to provided significant insight into design and manufacture vibration screen.

ACKNOWLEDGMENTS

The authors gratefully acknowledged the support from the National Natural Science Foundation of China (51175190).

REFERENCES

1. Chen YH, Tong X (2009) Application of the DEM to screening process: a 3D simulation. Min Sci Technol 19(4):493–497

2. Chen YH, Tong X (2010) Modeling screening efficiency with vibrational parameters based on DEM 3D simulation. Min Sci Technol 20(4):615–620

3. Chen QR, Yan F (1982) Theory and practice of the probability classification screen for coal. J China Univ Min Technol 2:25–33

4. Cleary PW (2004) Large scale industrial dem modeling. Eng Comput 21:169–204

5. Cleary PW (2015) Prediction of coupled particle and fluid flows using DEM and SPH. Miner Eng 73:85–99

6. Cleary PW, Dsinnott M (2015) Simulation of particle flows and breakage in crushers using DEM: part 1-compression crushers. Miner Eng 74:178–197

7. Cleary PW, Sawley M (2002) DEM modelling of industrial granular flows: 3D case studies and the effect of particle shape on hopper discharge. Appl Math Model 26(2):89–111

8. Dong KJ, Yu AB, Brake I (2009) DEM simulation of particle flow on a multi-deck banana screen. Miner Eng 22(11):910–920

9. Jiao HG, Ma J, Zhao YM, Chen LJ (2006) Study on the numerical simulation of batch sieving process. J Coal Sci Eng 12(2):80–83

10. Liu CS, Lu JX (2009) Study on nonlinear characteristic of particle motion during the process of coal screening. J China Coal Soc 34(4):556–559

11. Mindlin RD, Deresiewicz H (1953) Elastic sphere in contact under varying oblique forces. J Appl Mech 20(3):327–344

12. Soldinger M (2000) Influence of panicle size and bed thickness oil the screening process. Miner Eng 13(3):297–312

13. Soliding M (1999) Interrelation of stratification and passage in the screening process. Miner Eng 12(5):497–516

14. Subasinghe GKNS, Schaap W, Kelly EG (1989) Modeling the screening process: a probabilistic approach. Powder Technol 59(1):37–44

15. Wang GF, Tong X (2011) Screening efficiency and screen length of a linear vibrating screen using DEM 3D simulation. Ming Sci Technol 21(3):451–455

16. Weerasekara NS, Powell MS, Cleary PW, Tavares LM, Evertsson M, Morrison RD, Quist J, Carvalho RM (2013) The contribution of DEM to the science of comminution. Powder Technol 248:3–24

17. Wei LB, Zhao YM, Chen QR (1999) Study on the parameters by imitating the process of probability screen for coal sizing. J China Coal Soc 24(1):103–107

18. Wen BC, Liu FQ (1982) Vibration machine theory and application. China Machine Press, Beijing

19. Xiao JZ, Tong X (2012) Particles stratification and penetration of a linear vibrating screen by the discrete element method. Int J Min Sci Technol 22(3):357–362

20. Xue S, Yuan L, Wang JF, Wang Y, Xie J (2015) A coupled DEM and LBM model for simulation of outbursts of coal and gas. Int J Coal Sci Technol 2(1):22–29

21. Zhang EL, Tong X (2012) Study on loose mechanism of vibrating screen based on DEM. Min Process Equip 40(12):67–70

22. Zhao LL, Liu CS, Yan JX (2010) Min Sci Technol 20(2):276–280

23. Zhu HP, Zhou ZY, Yang RY, Yu AB (2008) Discrete particle simulation of particulate systems: a review of major applications and findings. Chem Eng Sci 63(23):5728–5770

9

CHAPTER

Mineralogy, Distribution, Occurrence and Removability of Trace Elements During the Coal Preparation of No. 6 Coal From Heidaigou Mine

Xiangfei Bai [1,2], Yue Wang[1,2] , Wenhua Li[3]

[1]*Beijing Research Institute of Coal Chemistry, China Coal Research Institute, Beijing 100013, China*

[2]*State Key Laboratory of Coal Mining and Clean Utilization, Beijing 100013, China*

[3]*National Institute of Clean-and-Low-Carbon Energy, Beijing 102209, China*

ABSTRACT

Optical microscopy and scanning electron microscopy in conjunction with energy dispersed X-ray spectrometry (SEM–EDX) were used to study the minerals and the concentrations of 33 trace elements in No. 6 coal from Heidaigou mine. The distributions, organic affinity and removability of 18 trace elements were studied by float-sink experiments. A determination of the maceral groups was also undertaken. A high mineral content, dominated by kaolinite, was found in No. 6 coal from Heidaigou mine. The bauxite content was relatively high and it was mainly present as individual particles in fusinite lumens or was intimately intergrown with carbonate minerals. The pyrite and quartz contents were low. Some marcasite with a parallel twin structure was observed by cross-polar reflected light. A small amount of bean-like goyazite was present in the calcite. The weighted trace element content in Heidaigou formations is relatively low, which is beneficial for coal processing and utilization.

The concentrations of Ga, Hg, Pb, Se, Th, Ta are relatively high compared with the average values of Chinese coals. As, Hg, Mo, Ge, Ga, Ta, Ti, W, Mn are mainly present in minerals while B, Be, Th, P, Sc, Sr, V, Y, Yb are mainly found in organic matter. As, Ge, Hg, Mo are mainly present in sulfides and Be, Th, P, Sc, Sr, Y, Yb are mainly present in inertinite. B and V are mainly present in vitrinite. The high organic affinity and the low theoretical removability of most trace elements cause difficulties in removing them during coal preparation.

Keywords: Trace elements Occurrence Organic affinity Removability Heidaigou coal

1 INTRODUCTION

The Heidaigou coal mine is located in the middle of the Junger coalfield in the Inner Mongolia Autonomous Region. More than 30 million tons of raw coal is produced every year. As a preferred feedstock for power plants Heidaigou coal is primarily used for the generation of electricity. A low degree of anogenic metamorphism is the main coalification type. The main mineable coal bed, the No. 6 coal seam, was formed during the drying terrestrial delta deposition environment of the late Carboniferous period (Mao and Xu 1999).

Research into minimizing the hazards caused by trace elements is an important part of coal utilization technologies. Studies into the distribution and the occurrence of trace elements in coal are important. Bench samples from a drill core of the No. 6 coal seam were used to study the distributions and occurrence of trace elements in Heidaigou coal. It was found that Ga, Se, Sr, Zr, Hg, Pb, Th and REEs were relatively abundant while the Ga concentration was sufficient for industrial use. Ga, Th and REEs were found to be mainly distributed in boehmite in the No. 6 coal (Dai et al. 2006a). Their relative abundance in boehmite comes from the periodic drying of peat because of changes in the water table during the coal forming period (Dai et al. 2007). It has been shown that high concentrations of Ga in weathered coal come from the absorption of humic acid (Wang et al. 2011). Lead and selenium are mainly present in galena, clausthalite, and selenio-galena while Hg is mainly found in clausthalite and sulfides. High concentrations of Zr are attributed to the presence of zircon and Sr is present in goyazite (Dai et al. 2006b, c; Li and Ren 2006). No. 6 coal from the Haerwusu surface mine that adjoins Heidaigou is enriched in Li, F, Ga, Se, Sr, Zr, REEs, Pb, and Th while Ga and F were mainly found in the clay minerals and partially in organic matter. Se and Pb are mainly present in the epigenetic clausthalite that fills the fractures. Sr and P may share the same carriers (Dai et al. 2008).

As a low-cost and mature technology coal preparation effectively reduces the amount of harmful elements in addition to improving the calorific values and lowering the ash and sulfur content of coal (Martinez-Tarazona et al. 1992). Coal preparation is thus believed to be an effective method to lower the concentrations of trace elements. This reduces the amount of environmental pollution by the trace elements (Pires et al. 1997; Klika et al. 1997, 2000; Tang et al. 2005; Song et al. 2010).

The geological setting of Heidaigou mine has been reported (Dai et al. 2006c). Production samples were used to study the coal's mineral composition. The distributions, occurrence and removability of trace elements from No. 6 coal were determined using float-sink experiments, petrological and statistical methods in this study. Correlation analysis was conducted to study the relationship among the trace elements and between the trace elements and the maceral groups, the ash content and the sulfur content. The theoretical content of trace elements in the macerals and minerals were then calculated using linear equations and the organic affinities of the trace elements are discussed. The removabilities of the trace elements were determined to provide guidance for the clean and environmentally-friendly use of Heidaigou coal.

2 EXPERIMENTS

2.1 Characteristics of the samples

The samples were obtained from the No. 6 coal seam at the Heidaigou surface mine according to GB/T 481-1993 (production coal sampling method). Proximate analyses (GB/T 212-2008), ultimate analyses (GB/T 476-2008,GB/T 19227-2008), total sulfur (GB/T 214-2007) and forms of sulfur (GB/T 215-2003), ash composition (GB/T 1574-2007), the fusibility of ash (GB/T 219-2008), the maceral groups (GB/T 8899-2008), the microlithotype composition (GB/T 15590-2008), and the mean maximum reflectance of the vitrinite (GB/T 6948-2008) were determined following Chinese standards and the results are listed in Tables 1, 2, 3, 4, 5 and 6.

Table 1. Results of the proximate and ultimate analyses (wt%)

Proximate analysis				Ultimate analysis				
M_{ad}	A_d	V_{daf}	FC_d	C_{daf}	H_{daf}	N_{daf}	O_{daf}	S_{daf}
8.65	24.67	38.50	46.33	81.52	4.51	1.30	12.02	0.65

Table 2. Total sulfur and forms of sulfur in the coal (wt%)

$S_{t,d}$	Forms of sulfur		
	$S_{p,d}$	$S_{s,d}$	$S_{o,d}$
0.49	0.33	0.00	0.16

Table 3. Ash composition analysis (wt%)

SiO_2	Al_2O_3	Fe_2O_3	CaO	MgO	SO_3	TiO_2	P_2O_5	Na_2O	K_2O
23.84	57.18	3.58	5.12	0.43	0.72	2.48	0.18	0.24	0.81

Table 4. Fusibility of the coal ash (°C)

DT	ST	HT	FT
1,380	>1,500	>1,500	>1,500

Table 5. Maceral groups and the mean maximum reflectance of vitrinite

Maceral group (vol%)				R^o_{max} (%)
Vitrinite	Inertinite	Liptinite	Mineral	
31.87	44.88	8.78	14.48	0.52

Table 6. Microlithotypes in Heidaigou coal (vol%)

Vitrite	Clarite	Inertrite	Vitrinertite	Trimacerite	Liptite	Durite	Minerite
11.33	4.30	17.38	12.11	13.48	0.00	29.88	11.52

Heidaigou coal is a low-rank bituminous coal with a high ash content. The sulfur content is low and dominated by pyritic sulfur. The Al_2O_3 content is high while SiO_2 is relatively low in the ash. This results a high ash fusion temperature. The mine is located at the north edge of the carboniferous-permian coal accumulation basin in Northern China. The vitrinite content is lower than the average of the area while the inertinite and liptinite contents are higher (Han 1996). The dominant vitrinite macerals are telocollinite and desmocollinite. A small amount of telinite is also present. The liptinite group is mainly represented by sporinite, which disperses in the desmocollinite and in the matrix of the inertinite in the bedding direction. Fragmental fusinite usually mixes with semifusinite. The main microlithotypes are vitrite, trimacerite and durite while a small amount of clarite, inertrite and vitrinertite are present. Vitrite is usually uniform while trimacerite is fragmental or of lineation shape. Durite is composed of an inertinite matrix and a clastic texture.

2.2 Mineral analysis

Minerals are the main carriers of most trace elements in coal. The distributions, occurrence and removability of the trace elements are largely affected by the presence of minerals. They are closely related to each other in terms of their genesis. Samples were prepared for microscopic analysis (ZEISS AXIOSKOP 40 optical microscope) according to the Chinese standard (GB/T 16773-2008). The optical characteristics of the minerals are described below and photographs were taken. A scanning electron microscope in conjunction with energy-dispersive X-ray spectrometry (SEM–EDS) was used to study the characteristics of the minerals and also to determine the distributions of elements in the macerals and in the minerals by spot and area scanning. Backscattered electron images (BSE) of the minerals were then obtained. The mineral compositions were determined by EDS both qualitatively and quantitatively. The acquisition time for the EDS analysis was 50 s and the scanned areas were determined by the size and morphology of the mineral particles.

2.3 Distributions and occurrence of the trace elements

2.3.1 Determination of the trace elements

The concentrations of Ag, As, B, Ba, Be, Cd, Co, Cr, Cs, Cu, F, Ga, Ge, Hg, Mn, Mo, Ni, P, Pb, Sb, Sc, Se, Sn, Sr, Ta, Th, Ti, U, V, W, Y, Yb, Zn in the No. 6 coal seam were determined.

The trace elements were determined by inductively coupled-plasma mass spectrometry (ICP-MS) except for As, F, Ge, Hg, Sb and Se. For ICP-MS analysis the samples were digested using a microwave high pressure reactor. F was determined by pyrohydrolysis in conjunction with a fluoride ion-selective electrode according to GB/T3558-1996. Hydride generation atomic fluorescence spectrometry analysis (HGAFS) was used to determine the concentrations of As, Ge, Sb and Se while cold-vapor atomic fluorescence spectrometry analysis (CVAFS) was used to determine the concentration of Hg.

2.3.2 Occurrence of the trace elements

The coal samples were crushed to less than 3 mm and fine particles of less than 0.5 mm were collected. Float-sink experiments were conducted using particles between 3 and 0.5 mm following the

Chinese standard GB/T478-2008. The ash and sulfur content, the maceral groups and minerals (divided into clay minerals, pyrite, quartz and carbonates) and the contents of As, B, Be, Ga, Ge, Hg, Mn, Mo, P, Sc, Sr, Ta, Th, Ti, V, W, Y, Yb were determined in products of different densities and in the fines of particle size less than 0.5 mm. The precision of the float-sink experiment and the determination of the trace elements were evaluated by balancing the trace element content between the raw coal and the weighted mean values in the products of different densities in the float-sink experiments and in the fines of particle size less than 0.5 mm. The calculation was conducted according to Eq. (1):

$$R = \sum_{j=1}^{n} C_j W_j / C \tag{1}$$

where, R is the ratio of the weight mean trace element content in the products of different densities and of that in raw coal (%), C is the content of trace elements in the raw coal (μg/g), C_j is the content of trace elements in the products of a certain density including the fines of particle size less than 0.5 mm (μg/g),W_j is the yield of products of certain density including the fines of particle size less than 0.5 mm (%).

The dominant mineral components are clay minerals and the pyrite content is low resulting in a simple mineral composition for the Heidaigou coal. The coal can be divided into vitrinite, inertinite and various minerals in terms of coal petrology while liptinite is ignored. The mass fraction of organic matter that consists of vitrinite and inertinite, the minerals in the raw coal and the products of different densities can be calculated according to Eqs. (2) and (3).

$$MM = 1.1A_{\mathrm{d}} + 0.5S_{\mathrm{p,d}} \tag{2}$$

$$ORG = 100 - MM \tag{3}$$

where, MM is the mineral content of the coal (mass fraction), ORG is the organic matter in the coal (mass fraction), A_{d} is the yield of ash in the coal (%), $S_{\mathrm{p,d}}$ is the pyritic sulfur content of the coal (%).

Because the macerals and the minerals in the coal petrological analysis are expressed as volume fractions the densities of vitrinite, inertinite and the minerals were 1.33, 1.40 and 2.60 g/cm^3, respectively, for the calculation of mass fraction in this paper.

A linear relationship was found for the trace element content of the raw coal (C_{m}) and for the different macerals and minerals (C_{mi}). The theoretical trace element (C_{mi}) content of the macerals and the minerals was calculated by multi-factor linear regression of the maceral and mineral compositions (W_{mi}). The trace element (C_{m}) content of the products of different densities were determined using Eq. (4):

$$C_{m} = \sum_{i=1}^{n} C_{mi} W_{mi} \tag{4}$$

where, C_{m} is the content of a specific element in the coal (μg/g), C_{mi} is the theoretical content of a specific element in vitrinite, inertinite and in the minerals (μg/g), and W_{mi} is the maceral composition of the products of different densities (%).

The amount of trace elements in the organic matter of the coal can be expressed by organic affinity. The organic affinity index (A_{o}) of different trace elements in the coal was calculated using Eq. (5):

$$A_o = C_o W_o / C \tag{5}$$

where, C_{o} is the average content of a specific trace element in the organic matter (μg/g), C is the content of trace elements in the raw coal (μg/g), and W_{o} is the content of organic matter in the coal (mass fraction).

2.4 Removability of the trace elements

The theoretical degree of trace element removal depends on the organic affinity and on fine coal recovery. It is the ratio of trace element content in waste coal to that in raw coal. Suitable separation density and trace element washability was obtained by float-sink experiments. The theoretical degree of trace element removal from the coal preparation can be calculated using Eq. (6).

$$\lambda = (C - C_r \times FCR)/C \tag{6}$$

where, λ is the theoretical degree of trace element removal (%), C is the trace element content of the raw coal (μg/g), C_r is the trace element content of the clean coal (μg/g), FCR is the fine coal recovery (%).

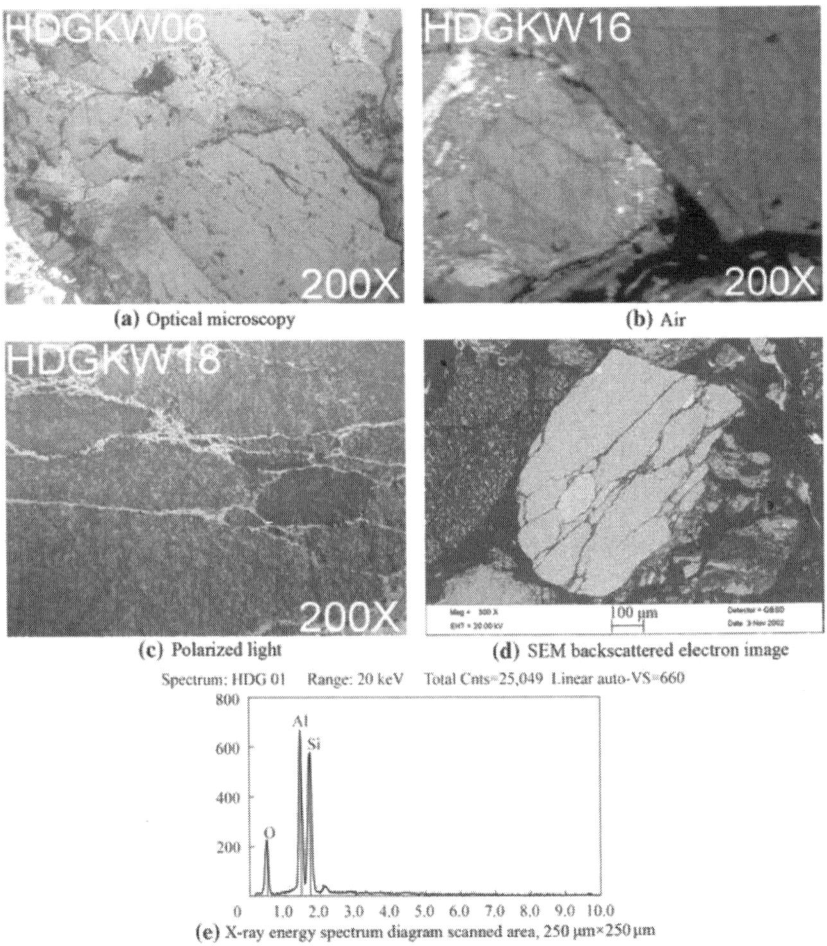

Figure 1. Kaolinite in Heidaigou coal

3 RESULTS AND DISCUSSIONS

3.1 Mineralogical composition of No. 6 coal from Heidaigou mine

A high mineral content, dominated by kaolinite, was found in No. 6 coal from Heidaigou mine. Its bauxite content is relatively high while pyrite and quartz are low.

1. *Clay minerals* Kaolinite is the main clay mineral in Heidaigou coal with some illite also present. Kaolinite is mainly dispersed in thin-layers or as individual particles (Fig. 1) and it is also found in the fusinite lumens. It appears uneven in reflected light with a gray–brown color and the kaolinite has a dim gray to yellow interference color under cross-polarized reflected light. The interference color changes slightly when the stage is rotated. The Si/Al ratio of Kaolinite is between 0.9 and 1.1 (Table 7) indicating a relatively high Al content and a comparatively low Si content. A low impurity content of zircon (Fig. 2), rutile (Fig. 3), chalcopyrite (Fig. 4) and Pb-bearing minerals (Fig. 5) was found in the Kaolinite by SEM–EDX analyses.

Table 7. Si/Al ratio of the clay minerals

Sample No.	Scanning area (μm × μm)	Acquisition time (s)	Si/Al ratio
1	250 × 250	50	1.04
2	250 × 250	50	1.06
3	100 × 80	50	0.90
4	250 × 250	50	1.01

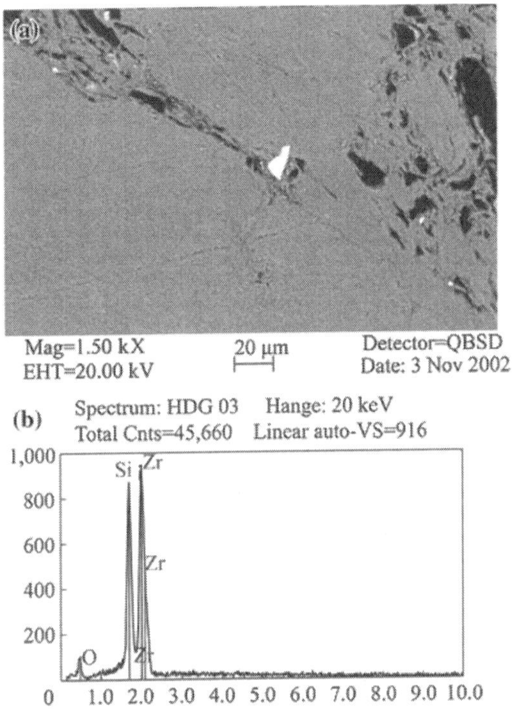

Figure 2. Fine-grain zircon in the kaolinite

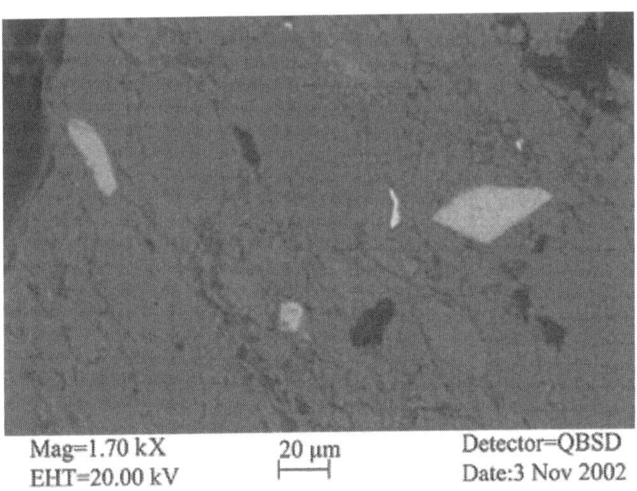

Figure 3. Fine-grain rutile in the kaolinite (SEM backscattered electron image)

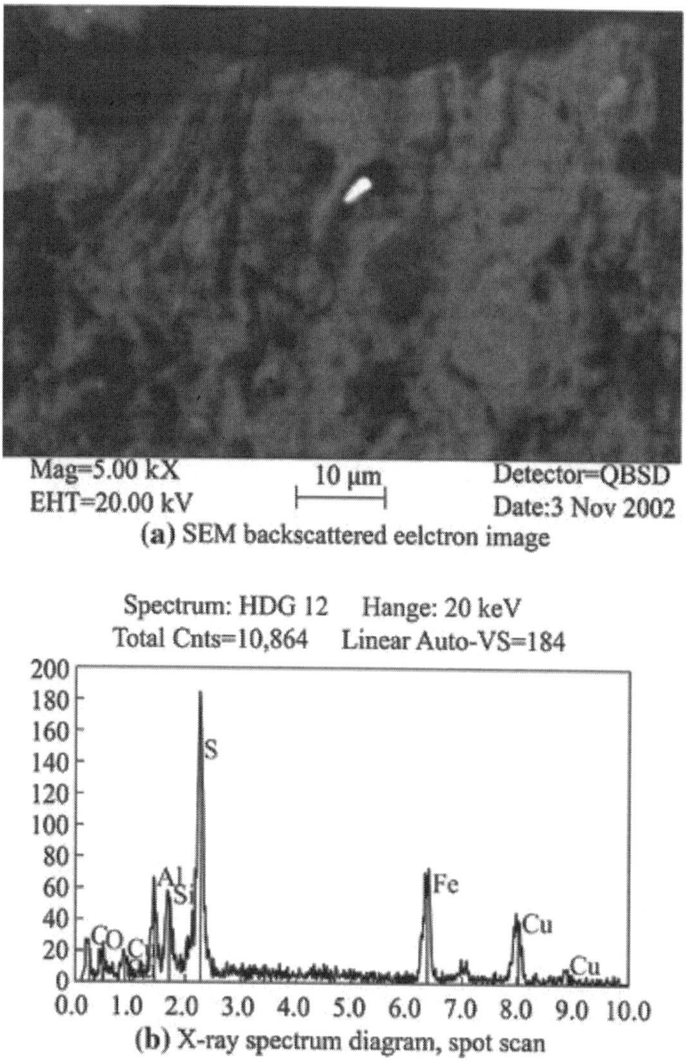

Figure 4. Fine grain chalcopyrite in the kaolinite

1. *Bauxite* The presence of bauxite in Heidaigou coal results in a relatively high Al_2O_3 content for the coal ash (57.18 %) and a high coal ash fusion temperature (temperature higher than 1,500 °C). A gray reflection color and an uneven surface were found for bauxite while a dim gray–yellow or no interference color was found when using cross-polarized reflected light. The interference color does not change when the stage is rotated. Therefore, it is difficult to distinguish bauxite from kaolinite

using optical microscopy. Individual particles (Fig. 6) are the main form of bauxite, which is also found in the fusinite lumens (Fig. 7) and it is intimately intergrown with carbonate minerals (Fig. 8).

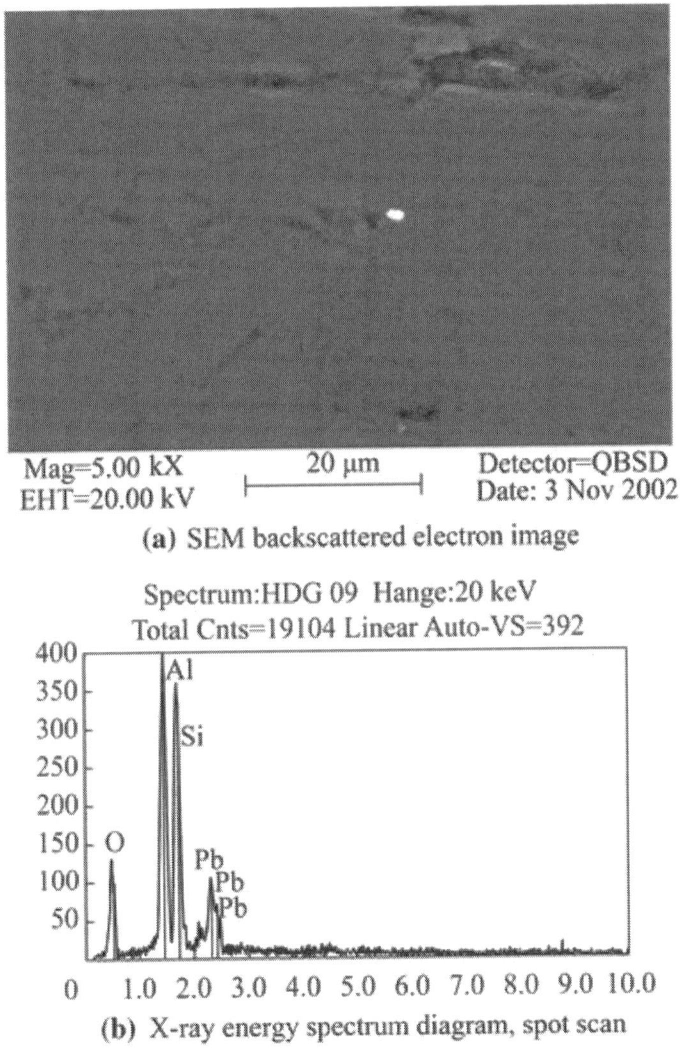

Figure 5. Pb-bearing minerals in the kaolinite

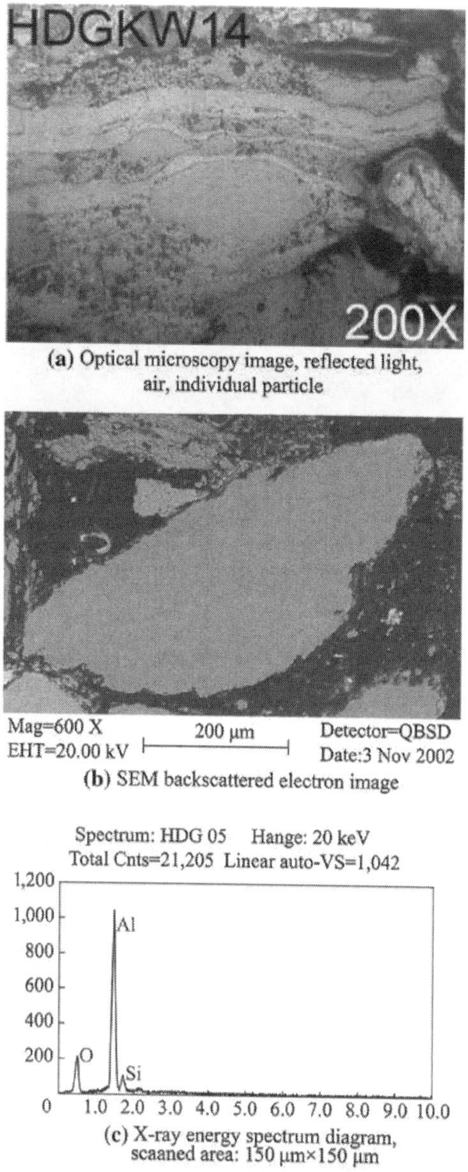

Figure 6. Bauxite in Heidaigou coal

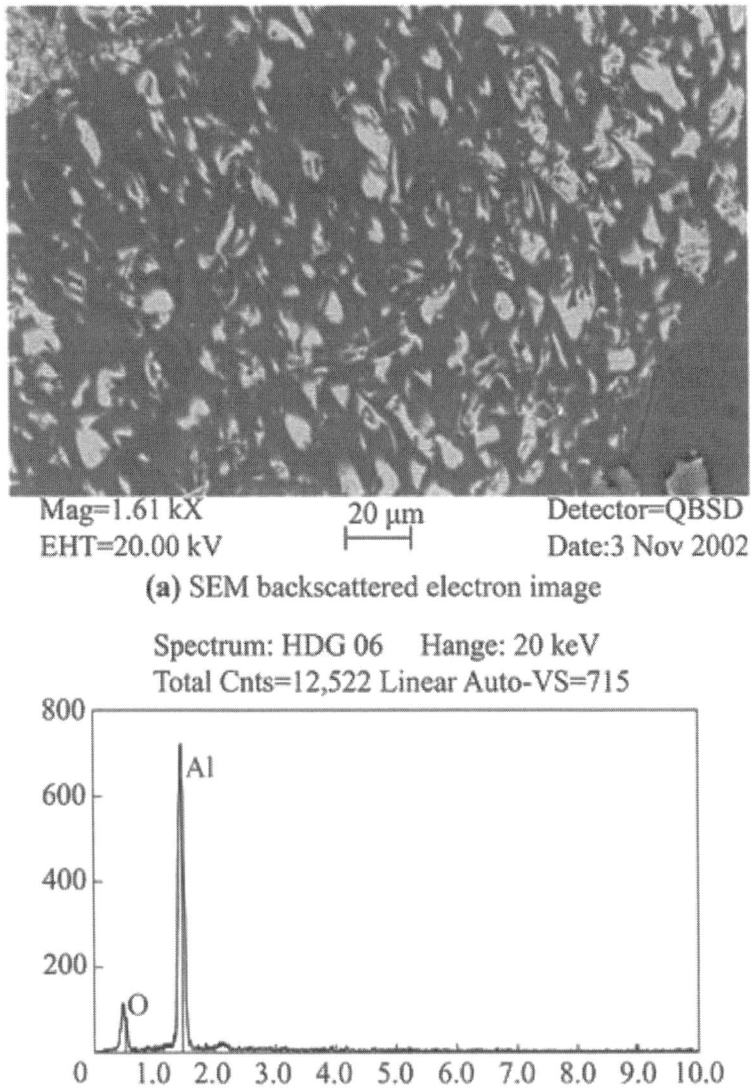

(b) X-ray energy spectrum diagram, spot scan

Figure 7. Bauxite in the fusinite lumens

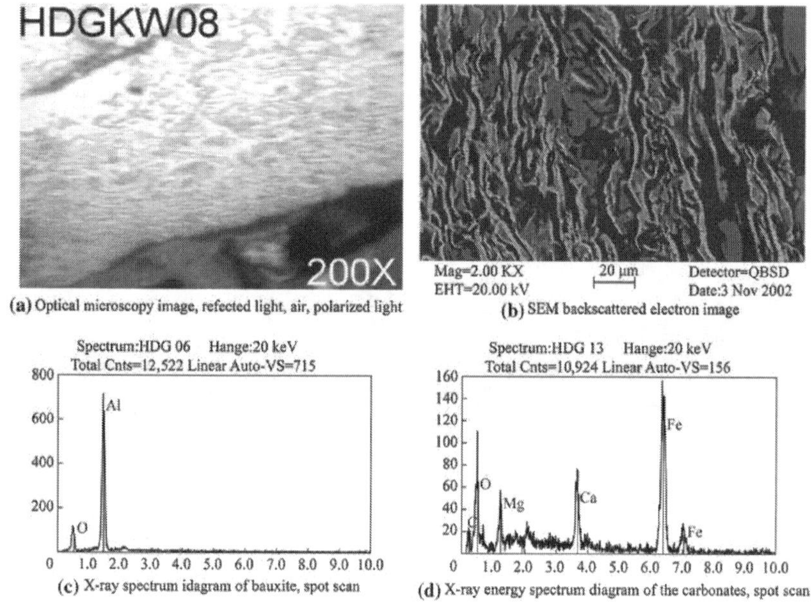

Figure 8. Bauxite intergrown with carbonates

1. *Carbonates* The carbonate mineral content is relatively high and the dominant mineral is calcite. Calcite mainly fills fractures (Fig. 9a) and cell cavities (Figs. 10, 11a, c). Some calcite was intimately intergrown with pyrite (Fig. 9b) and bauxite (Fig. 8). A significant amount of Fe and Mg that probably displaced Ca as isomorphs were found at the edge of the calcite (Fig. 11b, d). Small amounts of impure minerals such as irregular Al-bearing minerals and bean-like goyazite were found in the calcite particles (Fig. 11b, e).

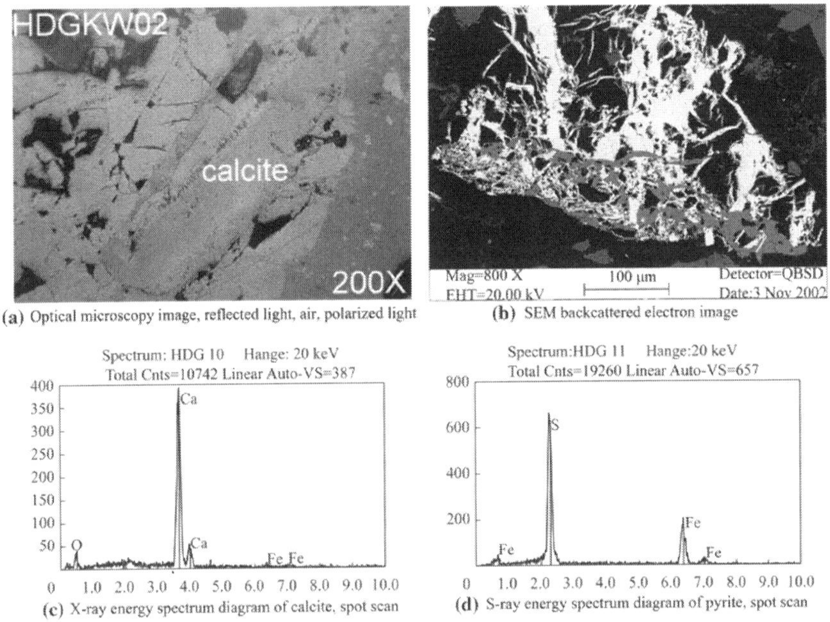

Figure 9. Pyrite and calcite that fill fractures

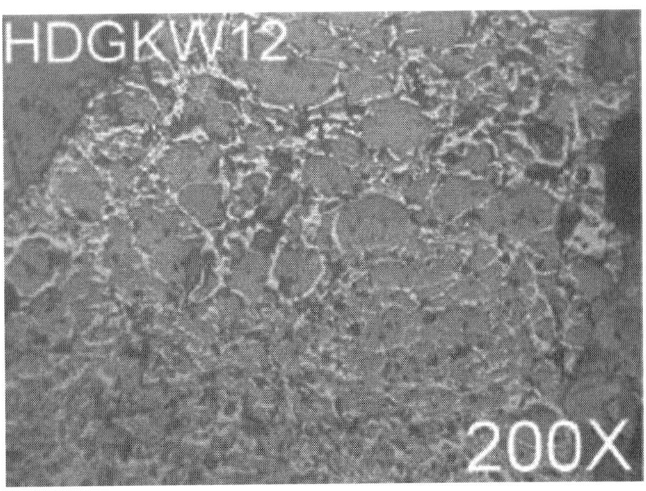

Figure . 10. Calcite in the fusinite lumens (reflected light; air; polarized light; ×200)

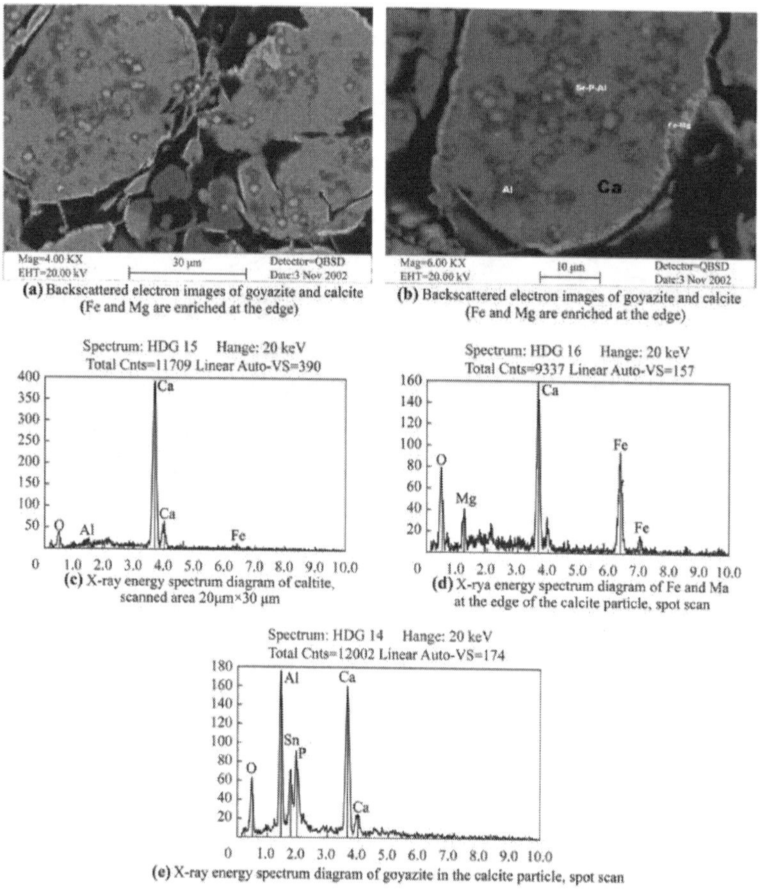

Figure 11. Calcite and goyazite (bean-like) in the fusinite lumens

1. *Sulfides* The sulfide content mainly comes from sulfides such as pyrite and it is low in coal. The pyrite mainly fills fractures (Fig. 9b) or disseminates in the fine grains. Some marcasite is present in the coal (Fig. 12). A similar yellowish-white reflected color was obtained for marcasite and pyrite when subjected to polarized light and it is thus difficult to distinguish marcasite from pyrite. Marcasite presents obvious anisotropy and it has a parallel twin structure and a dim gray-shallow celadon interference color in cross-polarized reflected light, which is significantly different to that of pyrite (Fig. 13).

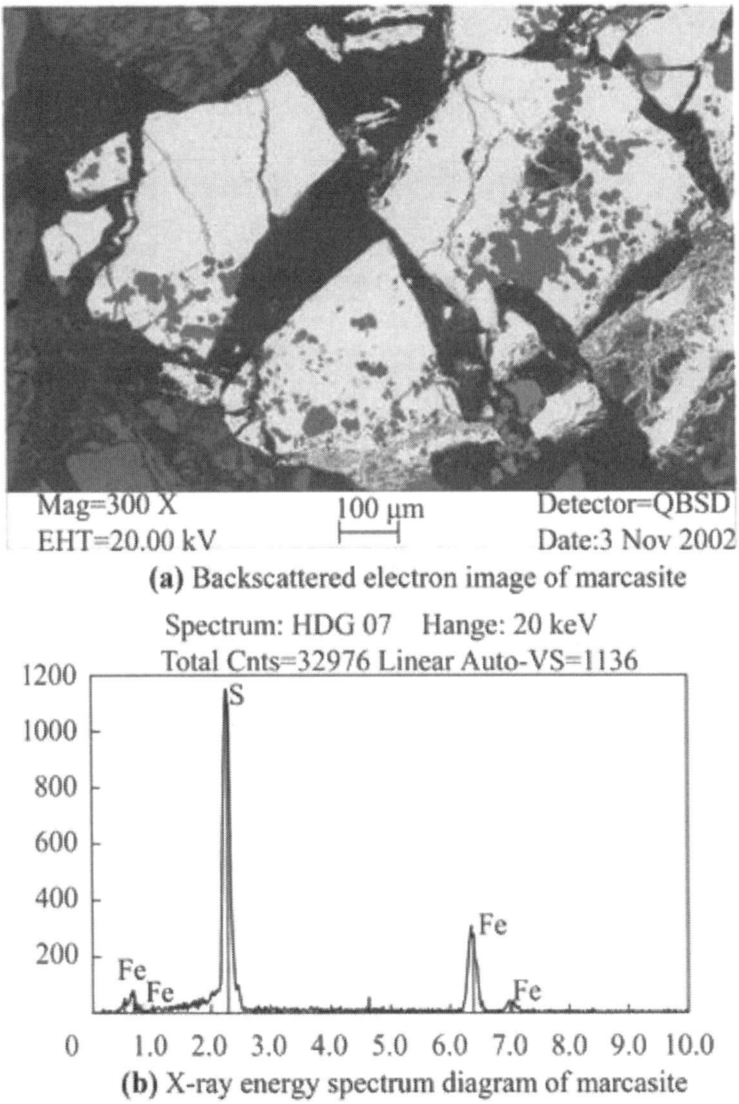

Figure 12. Marcasite in Heidaigou coal

Figure 13. Comparison between marcasite upon receiving reflected plane-polarized light and cross-polarized light

3.2 Distributions and occurrence of trace elements in No. 6 coal from Heidaigou mine

3.2.1 Trace element concentrations in the coal samples

The concentrations of 33 trace elements in No. 6 coal from Heidaigou mine were compared to average values of these trace elements in Chinese and international coals and the data are listed in Table 8. Compared to the average values of trace elements in Chinese coals, the weighted value of the trace elements in Heidaigou coal is relatively low. The concentrations of Ta, Ga, Hg, Th, Se, Pb, U and Ti are 1.5–3 times higher than the average of these elements in Chinese coals while the W, Y, P, Sr contents are almost the same as the average concentrations in Chinese coals. The concentrations of the other trace elements are lower than those in Chinese coals. Compared to the average values of trace elements in international coals the concentrations of Se, Ta, Ti, Th, Hg, Pb, Ga are 3–5 times

Table 8. Element concentrations in No. 6 coal from Heidaigou mine (μg/g)

Element	Concentrations	China coal*	World coal**
Ag	0.02	0.04	0.095
As	1.34	4.09	8.3
B	32.21	53****	52
Ba	18.1	270	150
Be	1.51	1.75	1.6
Cd	0.06	0.81	0.22
Co	1.24	10.62	5.1
Cr	3.00	16.94	16
Cs	0.10	1.51	1.0
Cu	12.53	17.87	16
F	100	157	88
Ga	19.06	6.84	5.8
Ge	0.09	2.43	2.2
Hg	0.413	0.154	0.10
Mn	40.9	271.22****	50
Mo	1.31	2.70	2.2
Ni	3.13	14.44	13
P	219.90	216	230
Pb	31.63	16.64	7.8
Sb	0.09	0.71	0.92
Sc	3.90	4.40	3.9
Se	7.09	2.82	1.3
Sn	0.51	2.11***	1.1
Sr	178.0	195	110
Ta	1.42	0.40	0.28
Th	15.31	5.88	3.3
Ti	2,515	1,685****	500
U	3.51	2.33	2.4
V	27.59	51.18	25
W	1.28	1.05	1.1
Y	10.68	9.07	8.4
Yb	1.58	1.76	1.0
Zn	18.0	41.4***	23

* Bai et al. (2007)
** Ketris and Yudovich (2009)
*** Dai et al. (2012)
**** Ren et al. (1999)

higher while Sr, Yb, and U are only slightly higher. Concentrations of Y, W, F, V, Sc, P, Be are almost the same as the average content of these elements in international coals. The concentrations of the other trace elements are lower than those in international coals. The ratios of these elemental concentrations in Heidaigou coal to Chinese coals and international coals are shown in Figs. 14 and 15.

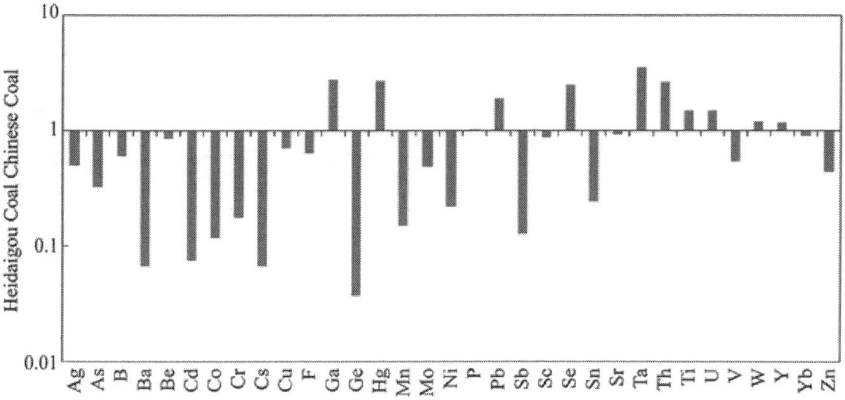

Figure 14. Comparisons between the concentrations of trace elements in Heidaigou coal and those in other Chinese coals

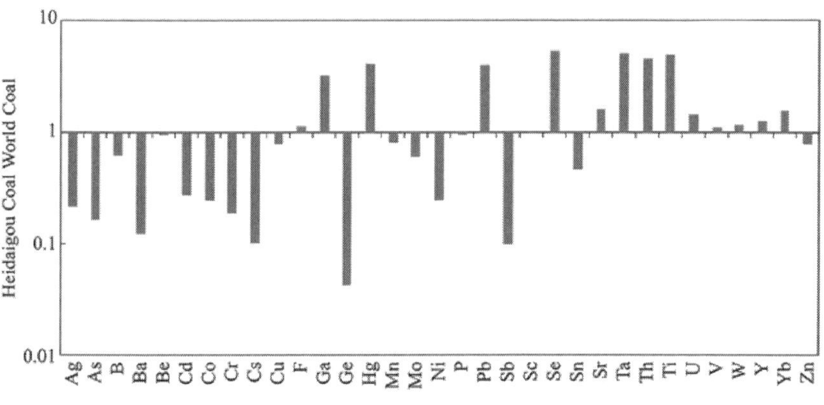

Fig. 15. Comparisons between the concentrations of trace elements in Heidaigou coal and those in international coals

Table 9. Maceral compositions and concentrations of 18 trace elements in the products from the float-sink experiments

Size grade (mm)	3–0.5						<0.5	Raw coal	*R* (%)
Yield of production (%)	54.44						45.56		
Density level (kg/L)	−1.30	1.3–1.4	1.4–1.5	1.5–1.6	1.6–1.8	+1.80			
Yield of products of different density (%)	6.29	31.12	19.82	13.34	8.46	20.97			
Coal quality									
A_d (%)	2.62	7.13	14.47	23.89	34.67	68.11	23.15	24.67	99.36
$S_{t,d}$ (%)	0.53	0.46	0.34	0.27	0.33	0.85	0.44	0.49	94.91
Maceral group (vov%)									
Vitrinite	79.09	40.54	14.21	10.56	8.53	5.99	27.62	31.87	80.13
Inertinite	12.55	40.54	65.94	63.34	56.81	22.66	50.90	44.88	105.63
Liptinite	7.27	13.21	5.10	3.84	0.91	0.75	4.69	8.78	63.56
Clay mineral	0.55	4.82	12.75	17.66	28.13	59.93	15.16	11.87	156.17
Pyrite	0.18	0.18	0.18	0.38	0.36	5.81	0.36	0.65	142.70
Quartz	0.18	0.18	1.28	1.54	2.54	2.25	0.36	0.33	249.84
Carbonates	0.18	0.54	0.55	2.69	2.72	2.62	0.90	1.63	72.81
Concentrations of trace element (μg/g)									

As	0.95	0.95	0.95	0.95	0.95	5.97	0.95	1.34	113.97
B	51.77	31.63	23.94	28.53	27.74	26.60	31.80	32.21	94.96
Be	1.11	1.63	2.10	2.03	1.94	0.83	1.63	1.51	106.88
Ga	14.73	14.76	15.27	19.86	25.05	27.10	18.99	19.06	99.65
Ge	0.19	0.13	0.06	0.06	0.06	0.24	0.08	0.09	117.71
Hg	0.063	0.091	0.113	0.155	0.175	0.971	0.342	0.412	76.62
Mn	10.00	13.10	24.60	42.30	48.60	41.70	45.80	40.90	88.39
Mo	1.32	1.05	0.98	1.10	1.37	2.02	1.21	1.31	95.72
P	145.50	253.90	315.00	262.70	200.10	97.57	222.80	219.90	101.38
Sc	2.19	4.28	5.40	5.11	3.75	0.95	5.48	3.90	116.20
Sr	152.40	248.40	258.60	204.00	111.10	41.73	203.60	178.00	108.24
Ta	0.32	0.58	1.02	1.29	1.91	2.51	1.39	1.42	93.24
Th	4.90	14.10	21.33	17.57	16.07	8.21	15.73	15.31	97.83
Ti	1,325.00	1,661.00	2,292.00	2,632.00	3,221.00	3,298.00	2,571.00	2,515.00	97.87
V	48.11	30.81	27.60	28.05	29.86	16.51	29.75	27.59	104.01
W	0.18	0.46	0.89	1.33	1.98	2.19	1.19	1.28	90.70
Y	12.93	15.99	17.08	13.72	9.49	3.53	14.21	10.68	124.58
Yb	1.51	1.80	2.06	1.79	1.45	0.53	1.74	1.58	103.09

The balance calculations for the maceral groups are based on volume fraction

3.2.2 Occurrence of trace elements in No. 6 coal from Heidaigou mine

The maceral composition and the content of 18 trace elements in the products from the float-sink experiments are listed in Table 9. The maceral compositions in the products of different densities are shown in Fig. 16. The precision of the float-sink experiments was in accordance with GB/T 478-2008, as shown by the balanced ash and sulfur content calculations. The *R* values of most trace elements were between 50 % and 150 %, which are acceptable for trace elements in float-sink experiments (Querol et al.2001; Reed et al. 2001). The dark reflectance colors of clay minerals and quartz make a quantitative determination in oil immersed reflected light difficult, which results in deviations in the results of the balance calculations.

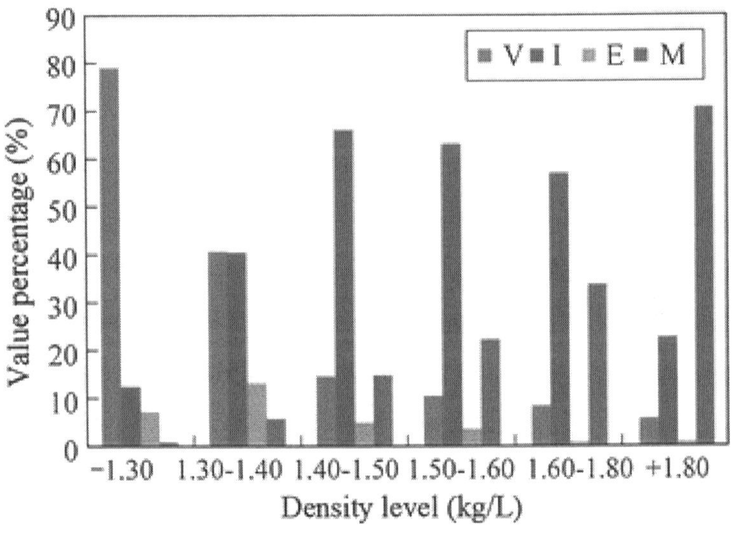

Figure 16. Histogram of the maceral groups in the products of different densities

As shown in Table 9: As, Ga, Hg, Mn, Mo, Ta, Ti, W concentrate in the heavy density products and, therefore, these trace elements are probably present in the minerals. B, P, Sr, V, Y, Yb are mainly present in the organic matter. B and V are mainly present in vitrinite while P, Sc, Sr, Y, Yb are present in inertinite. Be and Th are mainly present in inertinite with small amounts found in the minerals. Ge is evenly distributed in the minerals and in the organic matter.

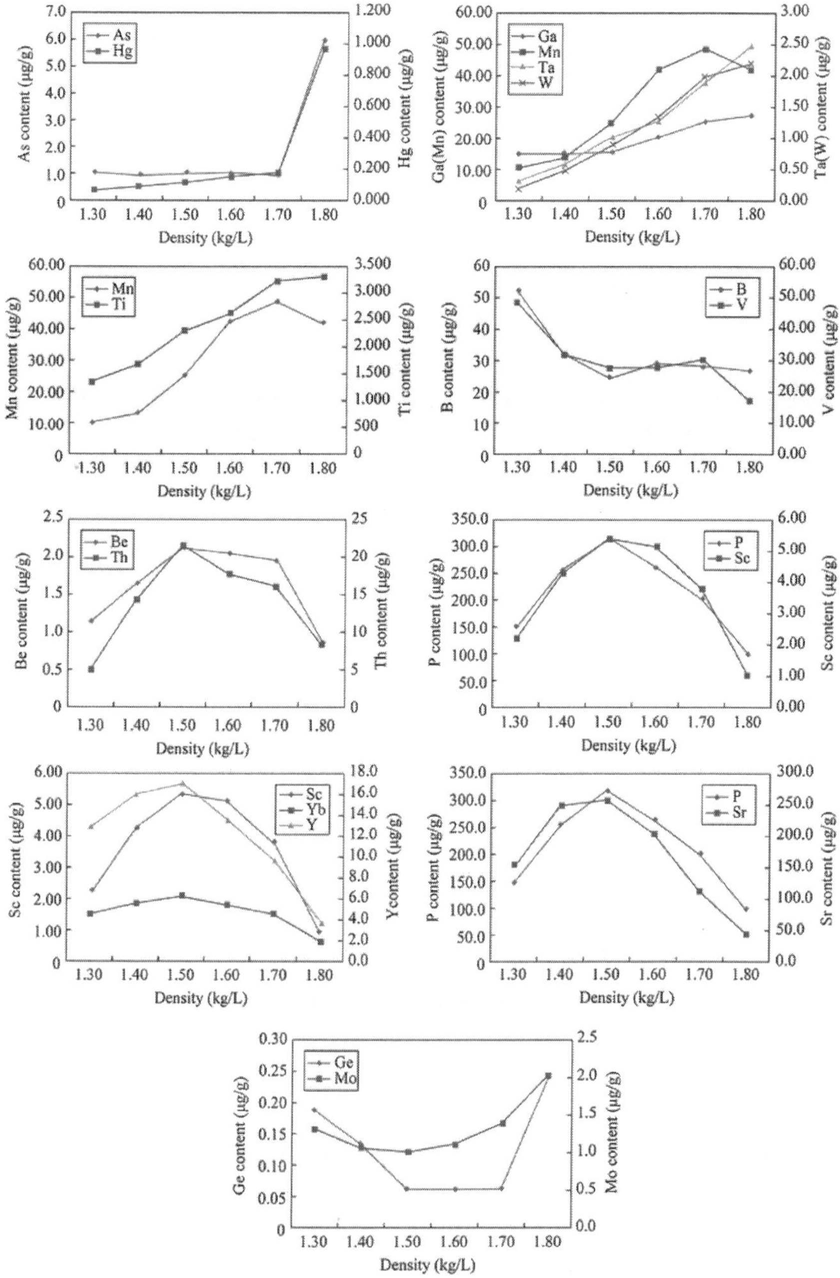

Figure 17. Trace element content of the products of different densities from No. 6 coal from Heidaigou mine

The trace element content of the products of different densities of the No. 6 coal from Heidaigou mine is shown in Fig. 17. The contents of As and Hg in the products of different densities have the same trend as the change in density. This is also the case for Ga, Ta, W, Ti and Mn. The trend for these elements is the same as that of the minerals upon a change in density. Therefore, these trace elements are probably present in the minerals. As the density changes the B and V content of the products of different densities have the same trend as that of vitrinite. Therefore, these trace elements are probably present in the vitrinite. The Be, Th, P, Sc, Yb, Y and Sr content of the products of different densities have the same trend as the change in density. This is the same as the behavior of inertinite. Therefore, these trace elements are probably present in the inertinite.

1. *Correlation analyses* The relationship between the trace element content and the maceral groups, the ash content, and the sulfur content are listed in Table 10 respectively. The content of trace elements, vitrinite, inertinite, clay minerals, pyrite, quartz, carbonates, ash and sulfur in the 6 products of different densities that were obtained from the float-sink experiments, raw coal and the fines of less than 0.5 mm were used to calculate the correlation coefficients. A significant positive correlation exists between As and the ash content, the sulfur content, the clay mineral content, and the pyrite at a confidence level of 95 % (critical correlation coefficient $R = 0.7068$). This is also the case for Hg and Mo. A significant positive correlation was found between Ga and the ash content, the clay mineral content, the quartz content and the carbonates content. It had a negative correlation with virtrinte and liptinite. This was also the case for Ta, Ti and W. A positive correlation was found between Mn and the carbonates content while a negative correlation exists between Mn and virtrinte. Significant negative correlations exist between Sr and the ash content, the sulfur content, the clay mineral content and the pyrite content. This was also the case for Y and Yb. A significant positive correlation existed between Be and inertinite and a negative correlation with the sulfur content. This was also the case for P, Sc and Th. A significant positive correlation existed between B and virtrinte and a negative correlation exists with inertinite. A

significant positive correlation exists between V and virtrinte and a negative correlation exists with the minerals. A significant positive correlation exists between Ge and the sulfur content. The correlation coefficients of the trace elements are listed in Table 11. A significant positive correlation exists among As, Hg, Mo and Ge. This is also the case among Ga, Ta, Ti, W, Mn, among Be, Th, P, Sc, Sr, Y, Yb and between B and V. High correlation coefficients indicate that these grouped elements have a similar occurrence. As, Hg, Mo, Ge, Ga, Ta, Ti, W, Mn are mainly concentrated in the minerals while Be, Th, P, Sc, Sr, Y, Yb, B, V are mainly found in the organic matter according to the correlation analyses of the trace elements and between the trace elements and the maceral groups.

Table 10. Relationship between trace element content and the maceral group content, the mineral content, the ash content and the sulfur content

Element	A_d	$S_{t,d}$	Vitrinite	Inertinite	Liptinite	Clay mineral	Pyrite	Quartz	Carbonates
As	0.8676	0.8810	−0.3512	−0.4,729	−0.4488	0.8905	0.9990	0.4708	0.4366
B	−0.5092	0.1370	0.9445	−0.7056	0.3505	−0.5078	−0.2594	−0.5762	−0.5163
Be	−0.3976	−0.9342	−0.3833	0.9369	0.0176	−0.3967	−0.6935	0.1181	0.0530
Ga	0.9323	0.4011	−0.6666	−0.0220	−0.7861	0.9056	0.7019	0.8163	0.8776
Ge	0.3744	0.9143	0.3432	−0.9200	−0.0179	0.4243	0.7266	−0.0155	−0.0705
Hg	0.9047	0.8304	−0.4204	−0.3560	−0.4554	0.8706	0.9392	0.3875	0.4869
Mn	0.6685	−0.0540	−0.7474	0.4421	−0.6863	0.5763	0.2755	0.5761	0.7903
Mo	0.8581	0.8630	−0.2093	−0.5941	−0.5389	0.8525	0.9184	0.4830	0.5038
P	−0.5626	−0.8124	−0.2050	0.8307	0.3495	−0.5597	−0.6967	−0.2154	−0.2674
Sc	−0.5142	−0.8395	−0.2097	0.8488	0.2303	−0.5431	−0.7325	−0.2555	−0.2096
Sr	−0.7936	−0.6761	0.1861	0.5303	0.6482	−0.7859	−0.7616	−0.5937	−0.6070
Ta	0.9636	0.4143	−0.7827	0.1020	−0.7567	0.9269	0.7338	0.7698	0.8166
Th	−0.1508	−0.7180	−0.6006	0.9756	−0.0224	−0.1736	−0.4434	0.1487	0.1118
Ti	0.8685	0.1581	−0.8751	0.3399	−0.7930	0.8275	0.5445	0.8169	0.8605
V	−0.7908	−0.3102	0.8760	−0.3295	0.3959	−0.7678	−0.6451	−0.5563	−0.6161
W	0.9155	0.2631	−0.8241	0.2135	−0.7847	0.8774	0.6260	0.8311	0.8875
Y	−0.8823	−0.7275	0.3110	0.4600	0.5839	−0.8518	−0.8419	−0.5803	−0.6737
Yb	−0.8354	−0.8647	0.1817	0.6264	0.4864	−0.8381	−0.9148	−0.4778	−0.5081

Table 11. Correlation coefficients for the trace elements in No. 6 coal from Heidaigou mine

	As	B	Be	Ga	Ge	Hg	Mn	Mo	P
As	1.0000								
B	−0.2441	1.0000							
Be	−0.7076	−0.4683	−1.0000						
Ga	0.6751	−0.4453	−0.2534	1.0000					
Ge	0.7448	0.4102	−0.9481	0.2056	1.0000				
Hg	0.9333	−0.3136	−0.6535	0.7304	0.5841	1.0000			
Mn	0.2380	−0.5743	0.1855	0.7763	−0.3424	0.4713	1.0000		
Mo	0.9135	−0.0241	−0.7859	0.7817	0.7459	0.8911	0.3393	1.0000	
P	−0.6954	−0.4054	0.8931	−0.5561	−0.8303	−0.6481	−0.0941	−0.8931	1.0000
Sc	−0.7446	−0.3534	0.8758	−0.4517	−0.9009	−0.5996	0.1326	−0.8568	0.9151
Sr	−0.7499	−0.0606	0.6605	−0.8360	−0.6126	−0.7218	−0.4080	−0.9413	0.8986
Ta	0.7099	−0.6181	−0.1930	0.9544	0.1326	0.8090	0.8206	0.7341	−0.4150
Th	−0.4555	−0.7285	0.9033	−0.1204	−0.8875	−0.3329	0.3456	−0.6225	0.8822
Ti	0.5146	−0.7107	0.−637	0.9164	−0.1225	0.6360	0.9053	0.5445	−0.2121
V	−0.6319	0.8771	−0.0125	−0.6477	0.0007	−0.7142	−0.6184	−0.4314	0.0028
W	0.5981	−0.6440	−0.0556	0.9620	0.0024	0.6957	0.8654	0.6463	−0.3345
Y	−0.8313	0.1007	0.6532	−0.8739	−0.5889	−0.8436	−0.4854	−0.9636	0.8363
Yb	−0.9107	0.0012	0.7972	−0.7671	−0.7791	−0.8612	−0.2830	−0.9921	0.9039
	Sc	Sr	Ta	Th	Ti	V	W	Y	Yb

As									
B									
Be									
Ga									
Ge									
Hg									
Mn									
Mo									
P									
Sc	1.0000								
Sr	0.8338	1.0000							
Ta	−0.3262	−0.7140	1.0000						
Th	0.8587	0.6142	0.0500	1.0000					
Ti	−0.1068	−0.5763	0.9640	0.2653	1.0000				
V	0.0285	0.2709	−0.8117	−0.3619	−0.7888	1.0000			
W	−0.2455	−0.6787	0.9829	0.1363	0.9885	−0.7744	1.0000		
Y	0.7964	0.9637	−0.8063	0.5133	−0.6593	0.4673	−0.7581	1.0000	
Yb	0.8845	0.9376	−0.6972	0.6595	−0.5025	0.4100	−0.6146	0.9572	1.0000

Theoretical trace element content in the maceral groups and the minerals in No. 6 coal from Heidaigou mine The trace element content of the maceral groups and of the minerals can be calculated using float-sink experiments and statistical methods, which is of great importance in determining the trace element occurrence and their behavior during coal processing and conversion. The theoretical content of trace elements in the maceral groups and in the minerals of No. 6 coal from Heidaigou mine are listed in Table 12. A ternary diagram showing the distribution of trace elements is given in Fig. 18. As, Ga, Hg, Mn, Mo, Ta, Ti, W are obviously concentrated in the minerals with low concentrations found in organic matter. B, Be, P, Sc, Sr, Th, V, Y, Yb are concentrated in the organic matter. B and V are mainly present in virtrinte while Be, P, Sc, Sr, Th, Y, Yb are present in inertinite. The Ge content of the minerals and of the organic matter is similar, which indicates an even distribution of Ge in the minerals and in the organic matter (no significant correlation from an F test).

Table 12. Theoretical trace element content in the maceral groups and in the minerals of No. 6 coal from Heidaigou mine, calculated values (in μg/g)

Element	Component			Sample amount	Significance testing			
	Vitrinite	Inertinite	Minerals		R	F	$F_{0.05}$ (3, $n-4$)	Significance
As	1.82	−3.01	7.01	8	0.93	10.73	6.59	Significant
B	54.45	16.34	29.29	8	0.94	12.61	6.59	Significant
Be	0.69	3.19	0.52	8	0.96	18.23	6.59	Significant
Ga	14.37	12.8	31.83	8	0.94	12.42	6.59	Significant
Ge	0.04	0.04	0.04	8	0.97	24.12	6.59	Significant
Hg	0.128	0.114	0.413	7	0.32	0.15	9.28	Non-significant
Mn	11.58	2.78	109.33	7	0.89	5.09	9.28	Non-significant
Mo	1.47	0.28	2.35	8	0.99	72.63	6.59	Significant
P	105.8	460.94	20.28	8	0.56	0.77	6.59	Non-significant
Sc	1.23	9.63	−0.56	8	0.96	22.45	6.59	Significant
Sr	138.63	388.46	−48.11	8	0.94	12.29	6.59	Significant
Ta	0.27	0.66	3.04	8	0.97	24.06	6.59	Significant
Th	1.17	32.64	4.01	8	0.99	82.17	6.59	Significant
Ti	1,082.59	2,404.66	3,738.02	8	0.96	17.89	6.59	Significant
V	46.93	27.65	16.46	8	0.88	5.86	6.59	Significant
W	0.03	0.81	2.71	8	0.96	17.64	6.59	Significant
Y	11.43	24.04	−1.27	8	0.95	15.28	6.59	Significant
Yb	1.26	3.03	0.06	8	0.99	137.42	6.59	Significant

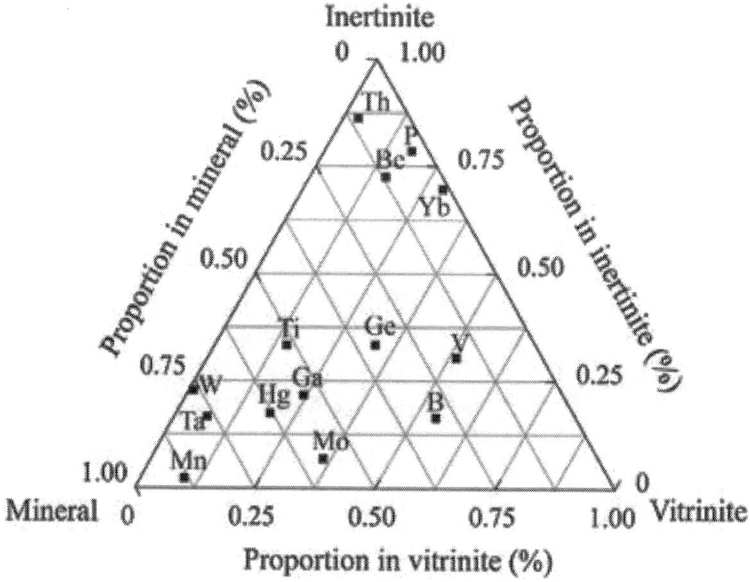

Figure 18. Ternary diagram showing the distribution of trace elements in No.6 coal

1. *Organic affinities of the trace elements in No. 6 coal from Heidaigou mine* The organic affinities of the trace elements in No. 6 coal from Heidaigou mine are listed in Table 13. The value for Ta in the raw coal was obtained from a weighted mean value of Ta in the different density products. As shown in Table 13, most of the trace elements are associated with the organics except for As, Hg, Mn, Ta, and W. Their low organic affinity indicates that these elements are mainly integrated into an inorganic phase. It is worth noting that trace elements with high organic affinity are not necessary organically integrated because some trace elements were present in the fine grained minerals in the organic matter that filled the lumens.

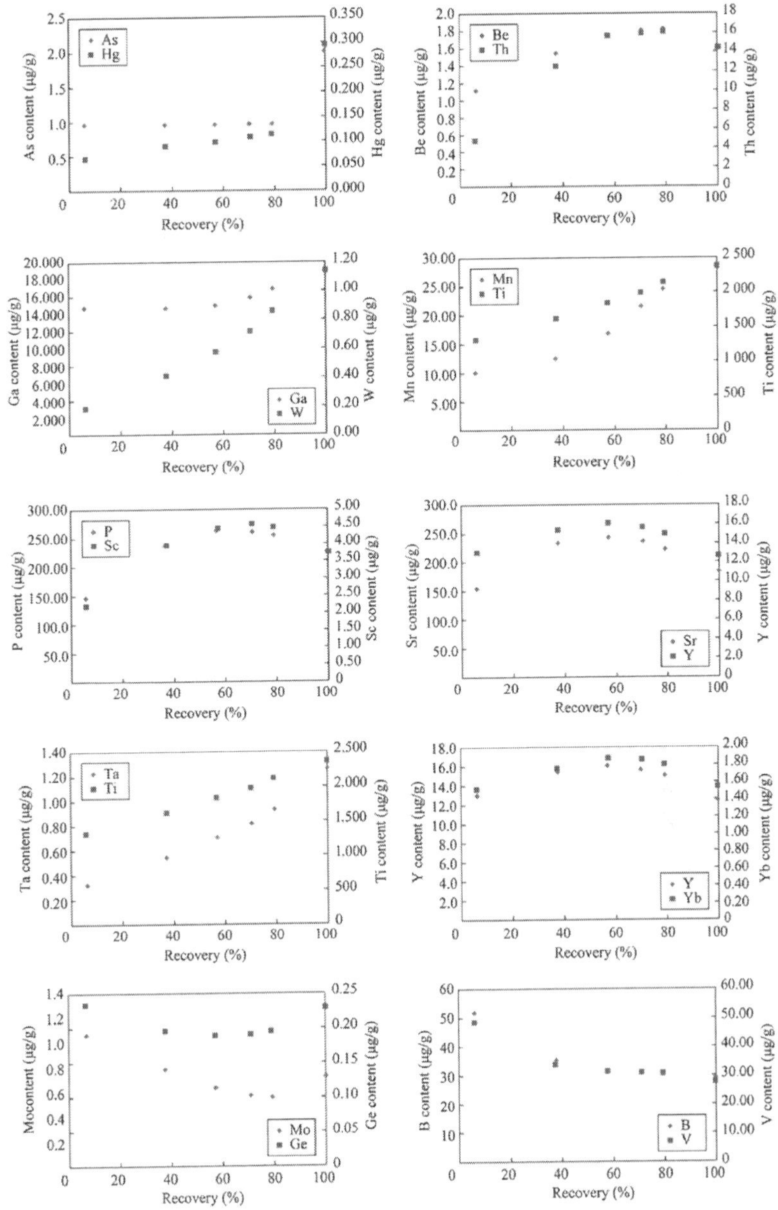

Figure 19. Washability curves for the trace elements in No. 6 coal from Heidaigou mine

Table 13. Organic affinity of the trace elements in No. 6 coal from Heidaigou mine

Element	As	B	Be	Ga	Ge	Hg	Mn	Mo	P
A_o	26.66	77.75	92.11	56.05	–	–	–	51.61	–
Element	Sc	Sr	Ta	Th	Ti	V	W	Y	Yb
A_o	100	100	32.05	93.07	58.93	86.98	33.05	100	99.10

3.3 Removal of trace elements during the coal preparation of No. 6 coal from Heidaigou mine

The washability curves of the trace elements were obtained during coal preparation of No. 6 coal from Heidaigou mine and are shown in Fig. 19. The theoretical removal of the trace elements is listed in Table 14. The value for Ta in the raw coal was obtained as a weighted mean value of Ta in the different density products.

Table 14. Theoretical removal of trace elements in No. 6 coal from Heidaigou mine (Recovery of clean coal was 57.23 % at a separation density of 1.50 kg/L)

Element	As	B	Be	Ga	Ge	Hg	Mn	Mo	P
Theoretical removal rate (%)	59.29	44.60	34.28	55.16	28.64	86.70	76.57	53.92	31.51
Element	Sc	Sr	Ta	Th	Ti	V	W	Y	Yb
Theoretical removal rate (%)	34.87	22.39	71.55	41.71	58.07	34.45	74.05	14.10	32.72

The washability curves of As and Hg are similar, which is also the case for Ga and W, Mn, Ta and Ti. High removal rates were obtained for these elements during coal preparation because sharp washability curves were obtained. The elements with flat washability curves are evenly distributed in the organic matter and in the minerals and they cannot be effectively removed during coal preparation.

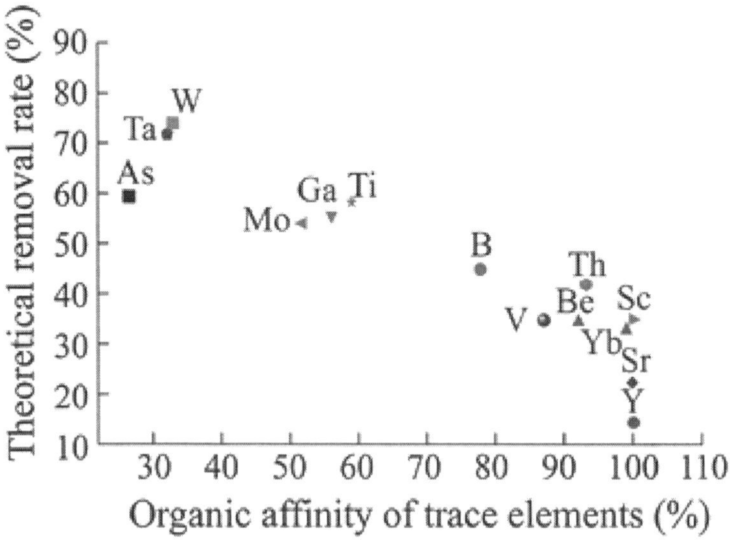

Figure 20. Relationship between organic affinity and theoretical trace element removal rate in No. 6 coal from Heidaigou mine

The ash and sulfur content of the clean coal were 9.18 % and 0.43 %, respectively, and poor washability was obtained for No. 6 coal. The recovery of trace elements is closely related to their washability, as shown in Table 14. Only Hg, Mn, W, Ta, As, Ti, Ga, Mo have high washability. The relationship between organic affinity and the theoretical removal rate of trace elements in No. 6 coal from Heidaigou mine are shown in Fig. 20. The theoretical removal rates of the trace elements decrease as the organic affinity increases. Most trace elements are closely related to the organic matter in No. 6 coal from Heidaigou mine. This is also shown by the correlation analyses and the theoretical concentration calculations of the trace elements in the maceral groups and in the minerals. Therefore, it is difficult to remove these trace elements during coal preparation.

4 CONCLUSIONS

Optical microscopy and SEM were used to study the characteristics of minerals and the concentrations of 33 trace elements in No. 6 coal from Heidaigou mine. The distribution of trace elements in the macerals and in the minerals as well as the organic affinity and removability of the trace elements are also discussed.

1. A high mineral content dominated by kaolinite clay minerals was found in No. 6 coal from Heidaigou mine. Bauxite was mainly present as individual particles in the fusinite lumens or was intimately intergrown with carbonate minerals. Its content was relatively high while that of pyrite and quartz was low. Some marcasite with a parallel twin structure, as observed by cross-polarized reflected light, was also found. A small proportion of bean-like goyazite was found in the calcite.

2. The weighted content of trace elements in Heidaigou formations is relatively low, which is beneficial for coal processing and utilization. The concentrations of Ga, Hg, Pb, Se, Th, Ta are relatively high compared to the average values of Chinese coals.

3. As, Hg, Mo, Ge, Ga, Ta, Ti, W, Mn are mainly concentrated in the minerals while B, Be, Th, P, Sc, Sr, V, Y, Yb are mainly found in the organic matter in No. 6 coal from Heidaigou mine. As, Ge, Hg, Mo are mainly present in sulfides, Be, Th, P, Sc, Sr, Y, Yb are mainly present in inertinite while B and V are mainly present in vitrinite.

4. The high organic affinity and low theoretical removability of most trace elements causes difficulties in their removal during coal preparation.

ACKNOWLEDGEMENTS

Our research was funded by the National Key Basic Research Development Plan (Grant Agreement number 2014CB744302).

REFERENCES

1. Bai XF, Li WH, Chen YF, Jiang Y (2007) The general distributions of trace elements in Chinese coals. Coal Qual Technol 1:1–4

2. Dai SF, Ren DY, Li SS, Chou CL (2006a) A discovery of extremely-enriched Boehmite from coal in the Junger coalfield, the Northeastern Ordos Basin. Acta Geol Sin 80(2):294–300

3. Dai SF, Ren DY, Li SS (2006b) Discovery of ultra-large gallium deposit in the Junger Coalfield Inner Mongolia. Chin Sci Bull 51(2):177–185

4. Dai SF, Ren DY, Chou CL, Li SS, Jiang YF (2006c) Mineralogy and geochemistry of the No. 6 Coal (Pennsylvanian) in the Junger Coalfield, Ordos Basin, China. Int J Coal Geol 66:253–270

5. Dai SF, Ren DY, Li SS, Zhao L, Zhang Y (2007) Succession characteristics of coal facies of the main coal seam in Heidaigou mine from the Jungar Coalfield, Inner Mongolia. China Earth Sci D 37(supp.):119–126

6. Dai SF, Li D, Chou CL, Zhao L, Zhang Y, Ren DY, Ma YW, Sun YY (2008) Mineralogy and geochemistry of boehmite-rich coals: new insights from the Haerwusu Surface Mine, Jungar Coalfield, Inner Mongolia, China. Int J Coal Geol 74:185–202

7. Dai SF, Ren DY, Chou CL, Finkelman RB, Seredin VV, Zhou YP (2012) Geochemistry of trace elements in Chinese coals: a review of abundances, genetic types, impacts on human health, and industrial utilization. Int J Coal Geol 94:3–21

8. Han DX (editor in chief) (1996). Coal petrology of China. China University of Mining and Technology Press, Xuzhou

9. Ketris MP, Yudovich YE (2009) Estimations of Clarkes for carbonaceous biolithes: world averages for trace element contents in black shales and coals. Int J Coal Geol 78:135–148

10. Klika Z, Kolonaznik I (2000) New concept for the calculation of the trace element affinity in coal. Fuel 79:659–670

11. Klika Z, Weiss Z, Roubicek V (1997) Calculation of element distributions between inorganic and organic parts of coal. Fuel 76(14/15):1537–1544

12. Li SS, Ren DY (2006) Analysis of anomalous high concentration of lead and selenium and their origin in the main minable coal seam in the Junger Coalfield. J China Univ Min Technol 35(5):612–616

13. Mao JH, Xu HL (1999) Prediction and evaluation of coal resources in China. Science Press, Beijing

14. Martinez-Tarazona MR, Spears DA, Tascon JMD (1992) Organic affinity of trace elements in Asturian bituminous coals. Fuel 71:909–917

15. Pires M, Fiedler H, Teixeira EC (1997) Geochemical distribution of trace elements in coal: modeling and environmental aspects. Fuel 76(14/15):1425–1437

16. Querol X, Klika Z, Weiss Z, Finkelman RB, Alastueya A, Juana R, López-Solera A, Plana F, Kolker A, Chenery SRN (2001) Determination of element affinities by density fractionation of bulk coal samples. Fuel 80:83–96

17. Reed GP, Ergüdenler A, Grace JR, Watkinson AP, Herod AA, Dugwell D, Kandiyoti R (2001) Control of gasifier mercury emissions in a hot gas filter: the effect of temperature. Fuel 80(5):623–634

18. Ren DY, Zhao FH, Wang YQ, Yang SJ (1999) Distributions of minor and trace elements in Chinese coals. Int J Coal Geol 40:109–118

19. Song DY, Zhang XK, Zhang JY, Zheng CG (2010) Migration characteristics of hazardous trace elements in coal in the process of flotation. J China Coal Soc 35(7):1170–1176

20. Tang YG, Chang CX, Zhang YZ (2005) Migration and distribution of fifteen toxic trace elements during the coal washing of the Kailuan Coalfield Hebei Province. Geochimica 34(4):366–372

21. Wang WF, Qin Y, Liu XH, Zhao JL, Wang JY, Wu DG, Liu JT (2011) Distribution, occurrence and enrichment causes of gallium in coals from the Jungar Coalfield Inner Mongolia. Sci China Earth Sci 41(2):181–196

10

CHAPTER

Modeling of Time Dependent Subsidence for Coal and ORE Deposits

Ryszard Hejmanowski[1]

AGH University of Science and Technology, Cracow, Poland

ABSTRACT

Coal and ore underground mining generates subsidence and deformation of the land surface. Those deformations may cause damage to buildings and infrastructures. The environmental impact of subsidence will not be accepted in the future by the society in many countries. Especially there, where the mining regions are densely urbanized, the acceptance of the ground deformations decreases every year. The only solution is to limit the subsidence or its impact on the infrastructure. The first is not rentable for the mining industry, the second depends on the precise subsidence prediction and good preventing management involved in the mining areas. The precision of the subsidence prediction depends strictly on the mathematical model of the deformation phenomenon and on the uncertainty of the input data. The subsidence prediction in the geological conditions of the raw materials used to be made on the basis of numerical modeling or the stochastic models. A modified solution of the stochastic model by Knothe will be presented in the paper. The author focuses on the precise description of the deposit shape and on the time dependent displacements of the rock mass. A two parameters' time function has been introduced in the algorithm.

Keywords: Subsidence prediction Influence function Modeling in time

1 INTRODUCTION

Specialists dealing with the protection of mining areas have recently discussed the issue of reliability and higher accuracy of deformation occurrence predictions (Kowalski 2005; Niedojadło 2008; Hejmanowski and Malinowska 2009). Such considerations stem from the increasing financial and social costs of damage done by this type of mining activity. The reliability issue is mainly connected with the quality of geologic-mining data (Naworyta and Sroka 2005; Stoch 2005), whereas higher accuracy of predictions may be related with the appropriate model of deposit and accuracy of the estimated parameters of the calculation model. Bearing in mind the popularity of Knothe theory and its common applicability in the Polish mining, the discussion may be focused on model parameters a and r (or interchangeably "tgβ"), and also additionally horizontal displacement parameter BB (Budryk 1953) and time coefficient c (Knothe1953). Over the years of study of deformations and predictions of their occurrence, the author noticed that deposit models used for calculations are frequently simplified (Knothe 2005), which results in considerable inaccuracies of the obtained results. The discrepancies of predictions and the actual deformations on the surface clearly indicate that such simplifications should never be made. The influence of a simplified description of deposit on the accuracy of model calculations is presented in this paper. An exemplary model employing author's prediction system MODEZ (Hejmanowski and Kwinta2009) is also presented. This program can be used for accurate modeling of deposition in a reservoir and the mining process itself.

2 CLASSIC MODELING OF DISPLACEMENTS IN A FUNCTION OF TIME

The most popular model of rock mass displacement in Poland and in many other countries all over the world is Knothe model worked out in 1953 (Knothe 1953), called the influence function model or stochastic model.

In this case, the vertical displacements of any point in the rock mass or on the surface can be modeled with this dependence:

$$S(x,y,z) = \frac{ag}{r_z^2} \iint \varphi(x,y) \mathrm{d}x\mathrm{d}y \tag{1}$$

where φ(x,y)φ(x,y) is the influence function in Knothe model, which will be simplified to the form: $\varphi(x,y) = \mathrm{e}^{-\pi\frac{x^2+y^2}{r_z^2}}$ for a point located at the origin of the coordinates system, *a* is the extraction coefficient which defines the way in which mining voids should be filled; *g* is the thickness of the deposit under production; and r_z is the radius of major influences in Knothe model, at a level *z* over the mined deposit. For *z* = *H*, this will be a dependence of the angle of major impact zone *β* and depth of extraction *H* (Fig. 1):

$$r(H) = H\mathrm{ctg}\beta \tag{2}$$

nd then

$$r_z = r_H \left(\frac{z}{H}\right)^n \tag{3}$$

where *H* is the depth of extraction (m); *β* is the angle of major impact zone (grad), and *n* is the rock mass parameter, characterizing propagation of deformations in the space above the deposit. The *n* value does not matter for the terrain surface, though according to many authors *n* < 1, 0 (Krzysztoń 1965; Drzęźla 1979; Preuße 1990).

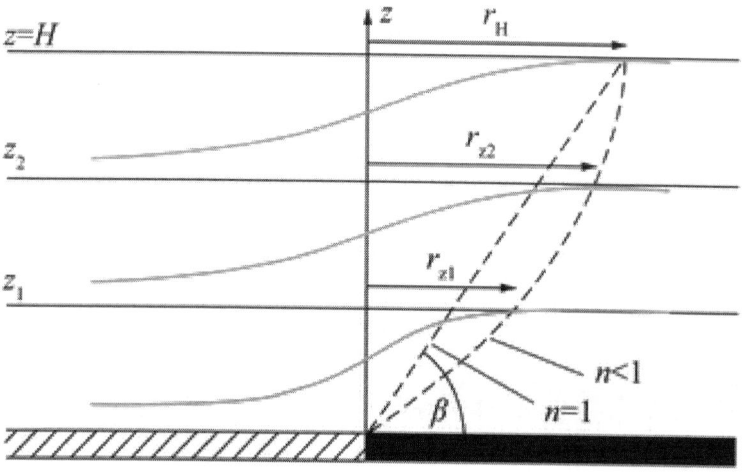

Figure 1. Mining impact dispersed in the rock mass

This model (1) was originally worked out for predicting static states of vertical displacements, however, later Knothe supplemented it by the function of time. This function was written in the form of a linear differential equation describing the subsidence rate as a function difference of final and momentary subsidence:

$$\frac{\delta S(t)}{\delta t} = c[S_K(t) - S(t)] \tag{4}$$

where $S_k(t)$ is the subsidence of a point when the mining has been stopped at a moment t; $S(t)$ is the actual subsidence of a point at a moment t, and c is the time parameter-proportionality coefficient in Eq. (4), characterizing delaying properties of the rock mass as far as mining impacts effects appearing in time [1/year] are concerned.

A general solution of (4) is (5):

$$S(t) = \exp(-ct) \int_0^t \exp(-c\tau) S_K(\tau) c\mathrm{d}\tau \tag{5}$$

This model well describes subsidence of the analyzed point of the terrain surface after the mining operations on a given mining field (or fragment of the lot) are finished. It can be hardly applied for the advancing longwall fronts, as in this case calculations should be made for fragments (strips) of the lot. The author of this model also noticed this regularity and suggested a solution for the advancing extraction front (Knothe 1984). A similar application of Knothe model for the calculation of dynamic subsidence of terrain was presented by a group of Chinese scientists (Hu et al. 2011).

If part of the deposit has been mined and depleted instantly (i.e. $S_K(t)$ = constants and $S(t)_{t=0} = 0$), the solution takes the form of Eq. (4):

$$S(t) = S_K(t)(1 - e^{-ct}) \quad (6)$$

As observed earlier (Kwinta et al. 1996), this function has a drawback, i.e. the subsidence is described from the moment of $t = 0$, when the subsidence rate is the maximum. This shortcoming can be liquidated by simulating the advancing extraction front in the form of strips (for longwall coal extraction), or deposit elements in the form of a cuboid.

3 NUMERICAL MODELING OF DISPLACEMENTS IN TIME-RESERVOIR ELEMENTS MODEL

This method was used in the calculation model based on the division of reservoir into reservoir elements (Hejmanowski 1993). In this case, the subsidence of an arbitrary point caused by mining activity can be written as:

$$\Delta S_{zi,j}(d_{i,j}, t_i) = V_i(t)\varphi_z(d_{i,j}, H_i) \quad (7)$$

and mining-induced subsidence of the entire mining lot or part of the reservoir is a sum of elementary impacts:

$$S_{zj}(t) = \sum_{i=1}^{N} \Delta S_{zi,j}(d_{i,j}, t_i) \quad (8)$$

where N is the number (quantity) of exploitation elements; i is the number of a subsequent individual exploitation element; j is the number of the (individual) calculation point; $V_i(t)$ is the volume of the post-exploitation elementary subsidence bowl at the time *t*, due to mining of the layer volume in the *i*th element of exploitation; H_i is the local mining depth in (m); *t* is the time difference between exploitation date of every reservoir element and date of the prediction (years); $d_{i,j}$ is the (horizontal) distance between the point calculation, and the centre of the exploitation element (m); and $\varphi_z(d_{i,j}, H_i)$ is the influence function, given with the formulae:

$$\varphi_z(d_{i,j}, H_i) = \frac{1}{r_z^2}\exp\left(-\pi\frac{d_{i,j}^2}{r_z^2}\right) \tag{9}$$

The model of reservoir division is presented in figure (Fig. 2). The horizontal dimension (square) is adjusted to the distance of the reservoir from the calculation level, accounting for the accuracy issue. The size of such elements is considerable, i.e. length *L* usually ranges from 30 to 60 m for extraction depth to 1000 m and predictions made for the surface. The height of the element corresponds with the thickness of the mined reservoir g_i.

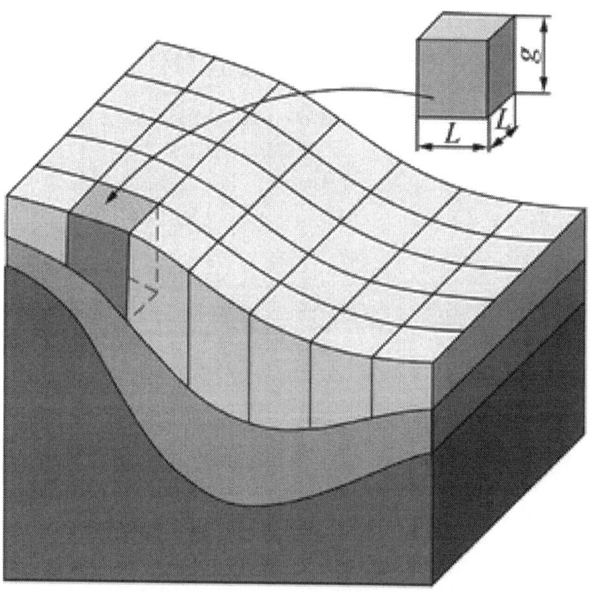

Figure 2. Exemplary division of reservoir into elements

3.1 Modeling of geology and mining technology

The division of reservoirs into elements has a very important utility aspect. All attributes connecting with geologic conditions and technological parameters of mining, which are locally variable, can be attributed to elements. In this situation the reservoir element may bring about information about the reservoir, which is especially significant for the prediction of deformation occurrence, and also spatial analyses and for evaluation of mining hazards.

3.1.1 Deposit thickness variable

The average thickness of extracted deposit layer is a typical generalization assumed in the traditional prediction models. It is not very important for coal seams, though in the metal ore deposits the thickness may vary significantly (Fig. 3).

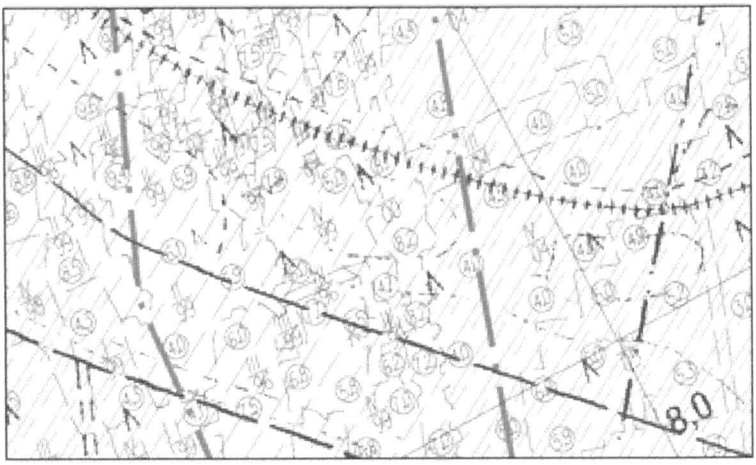

Figure 3. Exemplary thickness of a copper ore layer [thickness given in (m)]

The reservoir element model allows including this attribute in displacement calculations for thus diversified thicknesses of mined layers (Fig. 4). Consequently, the picture of the summary subsidence basin will vary, depending on the varying thickness of deposit.

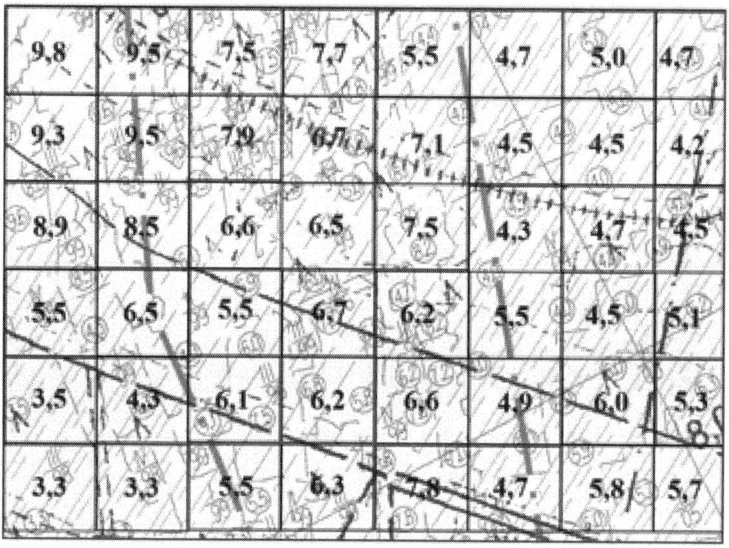

Figure 4. Reservoir elements with 'reservoir thickness g_i' attribute in (m)

3.1.2 Mixed mining system

The accuracy of displacement predictions is known to depend on the correct parametrization of the mathematical model. The parameters should correspond to the local geologic-mining conditions. The extraction coefficient a and angle of major impact β were the most important parameters in Knothe model, on the basis of which the presented modified model was worked out. In view of the planned extraction system, the extraction coefficient can also be simulated for the locally varying conditions, e.g. hydraulic backfilling (Fig. 5). In the presented example the room and pillar system with roof deflection is predicted for extraction coefficient of *ca.* 0.6. The system including hydraulic backfill better minimizes displacements and the extraction coefficient usually ranges between 0.2 and 0.3.

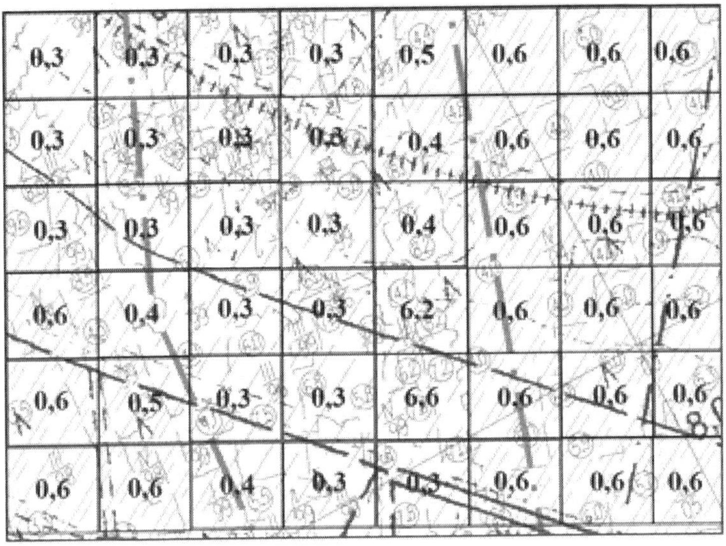

Figure 5. Reservoir elements with 'extraction coefficient a_i' attribute

3.1.3 Varying abundance of deposit and its modeling

The reservoir elements model allows for very accurate modeling of local changes in the ore resources thanks to which they can be selectively extracted. The influence of a given fragment of reservoir on the rock mass displacements can be modelled by introducing information about the surface of the planned mining area within the reservoir element as compared to the total reservoir element area. Figure 6illustrates a reservoir pillar which will not be extracted. Numbers corresponding to the percentage of mined area of a given reservoir element are introduced in particular cells of the model. Thanks to this the volume of an arbitrary reservoir element, i.e. its influence on rock mass displacement, can be calculated from dependence (10):

$$V_i^E = E_i \cdot L_i^2 \cdot g_i \tag{10}$$

where VEiViE is the volume of reservoir element to be mined; Ei is the Ei is the ratio of surface of mining area to total surface in a given reservoir element; LiLi is the length of the edge of reservoir element; and gigi is the local thickness of deposit layer to be mined.

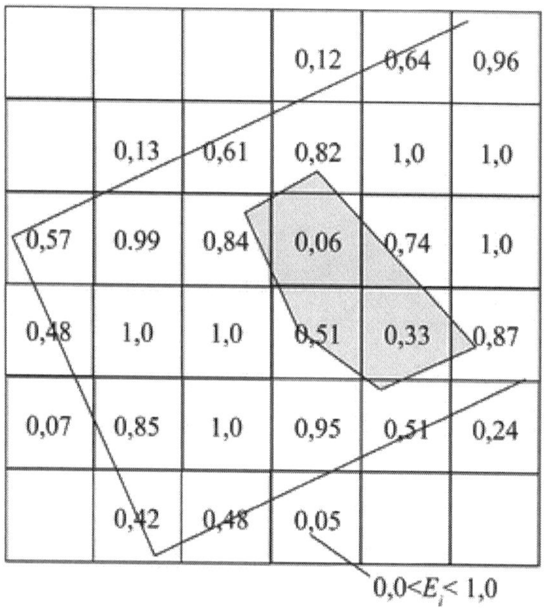

Figure 6. Modeling low-mineral pillar in mining area (excluded from predictions)

3.1.4 Tectonic disturbances and reservoir tilt

The biggest problem in modeling displacements is usually stratigraphy and tectonics of the reservoir formation. Unlike the presented model, classic models are frequently insufficiently flexible for modeling tectonic disturbances or variable reservoir tilt, e.g. within the coal longwall itself. Another advantage of the improved model is the local height datum as well as local direction and angle of dip of the deposit. Figure 7 reveals the way in which the difference of depth of deposition is accounted for. This affects the mining impact area of particular elementary basins on the surface. The summaric subsidence basin, being a sum of elementary basins, will show a visible asymmetry in the cross section parallel to the dip of deposit. This effect is observed in practice and no asymmetric influence functions need to be applied.

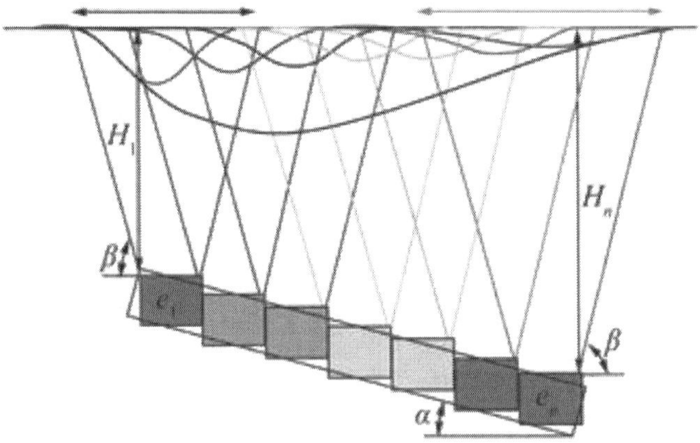

Figure 7. Asymmetry of summaric basin for inclined deposit layers

Additional information about the angle of inclination of deposit layers and the direction of the dip of the layer can be used for deviating the basin (m_i) i.e. shifting elementary basin towards the dip by a value calculated from dependence (11) (Fig. 8).

$$m_i = H_i \cdot \text{tg}(\mu\alpha) \tag{11}$$

where α is the angle of dip of deposit layer or roof formation (grad), $0 < \alpha < 50$ grad; μ is the deviation coefficient determined empirically for local geologic conditions, $\mu < 1$.

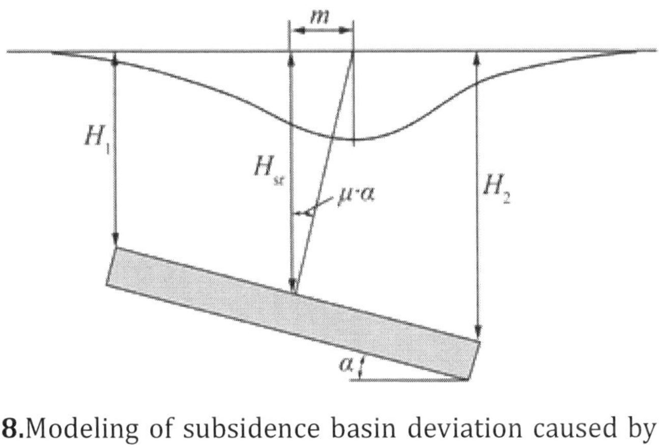

Figure 8. Modeling of subsidence basin deviation caused by deposit inclination

3.2 Time as important attribute of the model

Most importantly, the division of a reservoir into elements allows one to very accurately describe the mining development in time, and so to simulate the advancement of the extraction longwall or development of the room and pillar field, even for varying mining rates.

The extraction time of a given reservoir element as compared to the time, for which the subsidence has been calculated, is a unique pair of variables (dates) in the calculation algorithms. An arbitrary function of time can be used for this pair of variables to obtain a result which would be closest to the real conditions of mining deformation propagation in a given area. In this way the course of time in which manifests itself the mining impact at a given calculation point will automatically account for the space and time conditions referred to in the model.

It seems obvious that the surface deformation rate reaches the maximum after some time, but begins from the preliminary phase when the distance of the extraction from the given point on the surface is relatively big. Thus Knothe time model (4–6) will not be a suitable tool for describing this type of displacements. A far better solution will be the two-parameter function of time proposed by Sroka and Schober (Schober and Sroka 1983). This function, whose properties were discussed in detail (Kwinta et al. 1996) relates the cause of deformation, i.e. convergence of working with the subsidence basin. The volume of elementary basin needed in the forecast of displacements (7) is expressed with the formula (12).

$$V_i(t) = V_i^E \cdot \theta(t) \tag{12}$$

where $\theta(t)$ is the function of time.

$$\theta(t) = 1 + \frac{\xi}{\vartheta - \xi} e^{-\vartheta t} - \frac{\vartheta}{\vartheta - \xi} e^{-\xi t} \tag{13}$$

where ξ is the convergence coefficient characterizing delaying properties of roof rocks in the workings area, (year^{-1}); ϑ is the coefficient characterizing delaying properties of the cap rock in displacement propagation, (year^{-1}).

These two time parameters characterize time in which appear all mining impacts on the surface it can be assumed that they are the same as time coefficient introduced by Knothe. Then:

$$\frac{1}{c}=\frac{1}{\vartheta}+\frac{1}{\xi} \tag{14}$$

For traditional hard coal and ore extraction, the values of these coefficients take the following values:

(1)For hard coal (goaf system), $15 < \xi < 50$;

(2)For copper ores (room and pillar), $0.2 < \xi < 1.5$, $1 < \vartheta < 15$.

Graphic characteristics of the presented function of time are presented for exemplary values of time coefficient in figures (Figs. 9, 10).

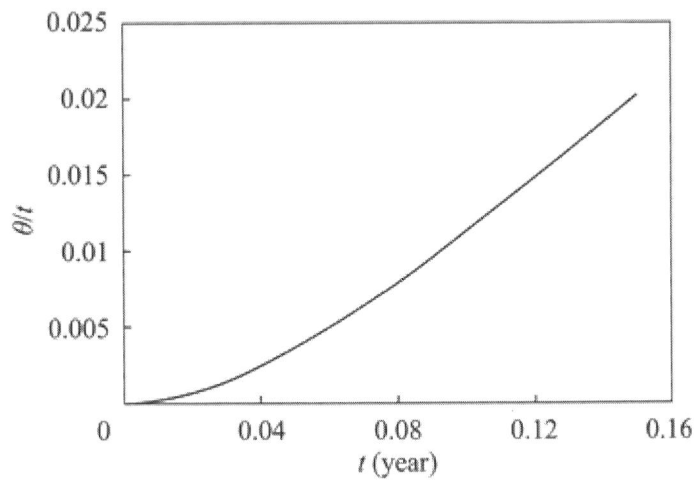

Figure 9. Schober-Sroka two-parameter function of time (0–0.15 year)

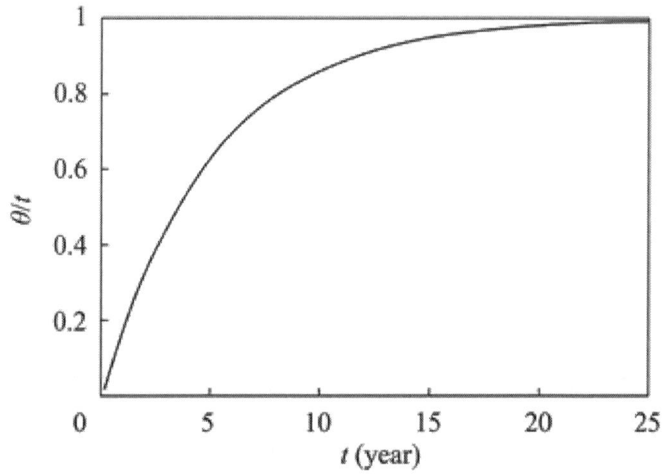

Figure 10. Schober-Sroka two-parameter function of time (0.15–25.0 years)

The function of time has zero rate and acceleration at the initial stage of movement. Then the displacement rate increases to the maximum in the linear phase of movement, delays, and the rate lowers at the vanishing movement stage. These properties of the function of time show to its adequacy to the actual conditions of displacement propagation in the rock mass as compared to Knothe function. Nonetheless the two-parameter function requires a much broader scope of studies to determine both these parameters. For the sake of determining the convergence coefficient, convergence measurements need to be performed in similar geologic and mining conditions, which is not always possible.

4 CONCLUDING REMARKS

The problem of modeling displacements in the coal or ore rock mass shows considerable advantages of a model based on Knothe influence function model. Modifications in the article in Knothe model refer to the space and time. The division of reservoir has the following advantages:

1) *(1)*this model gives much better possibility of including local variability of geologic properties of the rock mass and deposit as compared to Knothe model,

2) *(2)*properties characteristic of the designed extraction system (backfill, working convergence rate, leaving off pillars and unmined deposit remains) can be accounted for,

3) *(3)*time-modeling of extraction in planned reservoir lots, accounting for delays in displacements propagation in the rock mass.

Utilization of this model and author's computer software MODEZ (Hejmanowski and Kwinta 2009; Hejmanowski and Zespol (2009)) makes prediction of the rock mass deformations and displacements much easier in the coal and copper ore mining conditions. The introduction of this software to the mining practice will increase the accuracy of predictions.

REFERENCES

1. Budryk W (1953) Wyznaczanie wielkości poziomych odkształceń terenu. Archiwum Górnictwa i Hutnictwa, t.1, z.1, Warszawa

2. Drzęźla B (1979) Zmienność zasięgu wpływów w górotworze, Przegląd Górniczy, nr 10, Katowice

3. Hejmanowski R (1993) Zur Vorausberechnung förderbedingter Bodensenkungen über Erdöl- und Erdgaslagerstätten. Praca doktorska, TU Clausthal

4. Hejmanowski R, Zespół I (2009) Dokładność prognoz deformacji powierzchni terenu górniczego "Lubin I" i "Małomice I", w aspekcie zmian dynamiki ruchów powierzchni w odniesieniu do prognoz krótko- i długookresowych. Praca nauk-badawcza na zlecenie KGHM Polska Miedź SA. (niepublik.)

5. Hejmanowski R, Kwinta A (2009) System prognozowania deformacji Modez. Materiały konferencji X Dni Miernictwa Górniczego i Ochrony Terenów Górniczych", Kraków

6. Hejmanowski R, Malinowska A (2009) Evaluation of reliability of subsidence prediction based on spatial statistical analysis. Int J Rock Mech Min Sci 46:432–438

7. Hu QF, Cui XM, Wang G, Wang MR, Ji YX, Xue WW (2011) Key technology of predicting dynamic surface subsidence based on Knothe time function. J Softw 6(7):1273–1280

8. Knothe S (1953) Równanie profilu ostatecznie wykształconej niecki osiadania. Archiwum Górnictwa i Hutnictwa, t.1, z.1, Warszawa

9. Knothe S (1984) Prognozowanie wpływów eksploatacji górniczej. Wyd. Śląsk

10. Knothe S (2005) Asymetryczna funkcja rozkładu wpływów eksploatacji górniczej w ośrodku zmieniającym swoje własności. Archiwum Górnictwa 50(4):401–415

11. Kowalski A (2005) O błędach prognozy wskaźników deformacji powierzchni spowodowanych błędami parametrów teorii. Materiały konferencji, VIII Dni Miernictwa Górniczego i Ochrony Terenów Górniczych, Ustroń

12. Krzysztoń D (1965) Parametr zasięgu niecek osiadania w ośrodku sypkim, Archiwum Górnictwa, tom X, zesz.1, str. 29–51, Kraków

13. Kwinta A, Hejmanowski R, Sroka A (1996) A time function analysis used for the prediction of rock mass subsidence.//96 International symposium on mining science and technology, Xuzhou, pp 16–18

14. Naworyta W, Sroka A (2005) Ocena niepewności prognozy wskaźników deformacji terenu metodą symulacji geostatystycznej. Materiały konferencji, VIII Dni Miernictwa Górniczego i Ochrony Terenów Górniczych, Ustroń

15. Niedojadło Z (2008) Problematyka eksploatacji złoża miedzi z filarów ochronnych szybów w warunkach LGOM. Rozprawy i monografie nr 177. Wydawnictwo AGH, Kraków

16. Popiołek E, Ostrowski J (1981) Próba ustalenia głównych przyczyn rozbieżności prognozowanych i obserwowanych poeksploatacyjnych wskaźników deformacji. Ochrona Terenów Górniczych nr 58, Katowice

17. Preuße A (1990) Markscheiderische analyse und prognose der vertikalen beanspruchungvon Schachtsäulen im Einwirkungsbereich untertägigen Steinkohlenabbaus. Praca doktorska, TU Clausthal, Clausthal-Zellerfeld

18. Schober F, Sroka A (1983) Die Berechnung von Bodenbewegungen uber Kavernen unter Berucksichtigung des zeitlichen Konvergenz und Gebirsverhaltens. Kahli und Steinsaltz Bd.8, Heft 10 **(in German)**

19. Stoch T (2005) Wpływ warunków geologiczno-górniczych eksploatacji złoża na losowość procesu przemieszczeń i deformacji powierzchni terenu. Akademia Górniczo-Hutnicza w Krakowie, Kraków (praca doktorska)

11

CHAPTER

The Safety Journey: Using a Safety Maturity Model for Safety Planning and Assurance in the UK Coal Mining Industry

Patrick Foster [1,*] and Stuart Hoult [2]

[1]*Camborne School of Mines, University of Exeter, Cornwall Campus, Penryn, Cornwall TR10 9EZ, UK*

[2]*UK Coal Plc, Harworth Park, Blyth Road, Harworth, Doncaster, South Yorkshire DN11 8DB, UK*

ABSTRACT

A Safety Maturity Model was developed for use in UK coal mining operations in order to assess the level of compliance and effectiveness with a recently introduced standards based safety management system. The developed model allowed for a "self-assessment" of the maturity to be undertaken by teams from the individual sites. Assessments were undertaken at all sites (surface and underground) and in some cases within each site (e.g., underground operations, surface coal preparation plant). Once the level of maturity was established, improvement plans were developed to improve the maturity of individual standards that were weaker than the average and/or improve the maturity as a whole. The model was likened to a journey as there was a strong focus on continual improvement and effectiveness of the standards, rather than pure compliance. The model has been found to be a practical and useful tool by sites as a means of identifying strengths and weaknesses within their systems, and as a means of assurance with the safety management system standards.

Keywords: safety culture; safety management systems; safety maturity

1. INTRODUCTION

The maturity model concept is a recent research innovation within the discipline of safety management and has been applied to safety culture development within a number of "high hazard" industries such as the offshore [1,2], aviation [3], rail [4,5] and petro-chemical industries [6,7]. These models originated from both quality and organizational development models as well as capability maturity models that had been used in the software industry [7]. Such models were developed to allow organizations to understand their own level of safety culture maturity by assessing the level of compliance with various key elements of safety culture across a number of stages (typically 5) that represent different levels of maturity.

Industries where these models are well established, such as the oil and gas industry, are currently putting a considerable amount of effort into establishing best practice and identifying tools and techniques that are proven to be effective in improving safety [1]. Whilst this approach is sensible, its effectiveness may be limited by differences in the safety culture across and within the industry, and those companies/sites in the early stages of developing their culture will require different techniques that those with stronger cultures. Safety maturity models, which can be applied on a company by company, or site by site basis, can identify these differences and then help address any behavioral and cultural issues they may have with a view to improving the safety culture maturity over time.

2. SAFETY MATURITY MODELS

An example of a safety maturity model is given by Fleming and Lardner [8] who show a three stage model developed by an offshore operating company. The three stages of safety culture in this model are (i) dependent, (ii) independent and (iii) interdependent. In a "dependent culture" there is an emphasis on management and supervisory control, with a heavy focus on written rules and procedures. Their study [8] states that if an organization with such a culture wants to improve its maturity it needs to develop an "independent culture" where the focus is on a personal commitment to and responsibility for safety. Whilst there will still be rules and standards, employees look after their own safety. The final stage is "interdependent" where there is a team commitment to safety with everyone having a sense of responsibility for safety beyond their own work and by caring for the safety of others. In this model, the drivers of the safety culture improvement process are teamwork, trust and the ownership of safety by the workforce. The same stages and drivers are defined by DuPont's "Bradley Curve".

Two of the more common cited maturity models are those subsequently developed by Fleming [1,6] and that of Hudson [2,9,10]. Fleming's model [1] looks at the level of maturity across 10 elements of safety culture adapted from those listed by the UK Health and Safety Executive in their human error guidance, HSG48 [11] namely:

- Management commitment and visibility;
- Communications;
- Production *versus* Safety;
- Learning organization;
- Safety resources;
- Participation;
- Shaded perceptions about safety;
- Trust;

- Individual relations and job satisfaction.

The five levels of maturity are set out as a number of iterative stages that an organisation can move through. These stages of maturity are (i) Emerging, (ii) Managing, (iii) Involving, (iv) Co-operating and (v) Continually Improving; and the model provides a framework to assist in the selection and implementation of appropriate behavioral interventions. The overall level of safety culture maturity is determined by taking a mean of the level of maturity at each of these individual ten elements. The model suggests that a stage by stage approach in improving safety culture is required and that skipping a stage in the sequence is difficult [1]. It is likely that a site will spend some time, perhaps years, successfully progressing from one stage to the next. This safety culture improvement process is undertaken by sites assessing their current level of maturity and then by developing and implementing a plan to move to the next level based on removing the weaknesses of the previous level [12]. The implementation of this plan is monitored and the level of maturity re-assessed to evaluate success and any further actions.

The maturity refers to the maturity of the organization's behaviors and not the maturity of the safety management systems. Therefore Fleming [1] also states that his model is only really of relevance to organizations where the technical and systems aspects of safety are performing adequately and the majority of accidents that occur appear to be due to behavioral or cultural factors. Lardner [6] gives a case study of the use of this model at a large petrochemical plant in the UK where the maturity of most of the individual safety culture elements were at level 3 (Involving) moving to level 4 (Co-operating) indicating a mature culture but wary of some differences in culture across different occupational groups.

Hudson's Model [2,9,10] has been used to describe the safety maturity in many industries including oil, aviation and healthcare. Hudson's model illustrates a five step progression from a "pathological" stage where there is a "no care" culture and "no systems" through to a "generative" stage where managing risks is a way of life and fully integrated systems are effectively in place. The descriptions of each stage of safety culture are given below [7,9].

- Pathological—Safety is a problem caused by workers. The main drivers are the business and a desire not to get caught by the regulator.
- Reactive—Organisations start to take safety seriously but there is only action after incidents.
- Calculative—Safety is driven by management systems, with much collection of data. Safety is still primarily driven by management and imposed rather than looked for by the workforce.
- Proactive—With improved performance, the unexpected is a challenge. Workforce involvement starts to move the initiative away from a purely top down approach.
- Generative—There is active participation at all levels. Safety is perceived to be an inherent part of the business. Organisations are characterised by chronic unease as a counter to complacency.

Parker *et al.* [10] developed a framework consisting of 18 elements, and developed a question set relating to each element and each maturity level. This gave rise to a tool that could be used by organizations to understand their safety culture maturity. These eighteen elements were classed as either concrete elements [*i.e.*, relating to the safety management system (11 elements in total)] or abstract elements [*i.e.*, relating to the attitudes and behaviors (7 elements in total)]. Therefore this model could be applied to organizations with weaker safety management systems unlike Flemings' model [1]. The eighteen elements used are given in Table 1.

3. THE USE OF SAFETY MATURITY MODELS IN THE MINING INDUSTRY

Within the mining industry there are many differences in safety culture and the state of safety management across operations which will limit the effectiveness of best practice tools and techniques that companies are trying to introduce. Like the oil industry, sites at the early stages of safety maturity will require different techniques that those with stronger maturities. Mining is therefore well suited to the use of maturity models and indeed, some have been developed and used.

Table 1. Elements of safety culture within Hudson maturity model [10].

"Concrete" Elements	"Abstract" Elements
Benchmarking, Trends & Statistics	Who causes accidents in the eyes of management?
Audits & Reviews	What happens after an accident? Is the feedback Loop being closed?
Incident/Accident Reporting, Investigation & Analysis;	How do safety meetings feel?
Hazard and Unsafe Act reports	Balance between HSE & Profitability?
Work planning including PTW, Journey Management	Is management interested in communicating HSE issues with the workforce?
Contractor Management	Commitment level of the workforce and level of care for colleagues.
Competency/Training	What is the purpose of procedures?
Work-site Job Safety Techniques	
Who Checks Safety on a day to day basis?	
What is size & status of the HSE Department?	
What are the rewards of good safety performance?	

A team from the University of Queensland developed the MIRM (Minerals Industry Risk Management) Maturity Chart that was based on both the Hudson model and a similar approach used by Bayside Aluminum, a BHP Billiton site in Richards Bay, South Africa [13,14,15]. The MIRM model is described as a ladder that has five stages or "rungs". The terms used are slightly different from those used by Hudson [2,9] although the implied levels of maturity are similar. The lowest rung is called "Vulnerable" where the site will "accept that accidents happen". The next rung or level of improvement sees the site move to "Reactive" where there is recognition that the site should "prevent a similar incident". Improvement from this rung moves the site to "Compliant" where the culture and systems try to "prevent incidents before they occur". The next rung in the ladder is probably the largest or most profound step for any site. Movement to the "Proactive" rung involves the site, through its culture and methods, embracing the systems approach. At this rung of the ladder the system ownership genuinely moves to line management and supervision. The final rung in the MIRM ladder is titled "Resilient" which is used to describe a site that has successfully integrated safety and risk management into its operations. The rungs are shown in Figure 1, along with the specific characteristics expected at each stage.

The MIRM ladder was not written as a guideline but rather a description of a journey that is intended to fit any minerals industry site. It has been provided to assist sites with identifying their current status on this journey and potential next steps for improvement. Like the Hudson model there is a clear distinction between the cultural elements and the system elements.

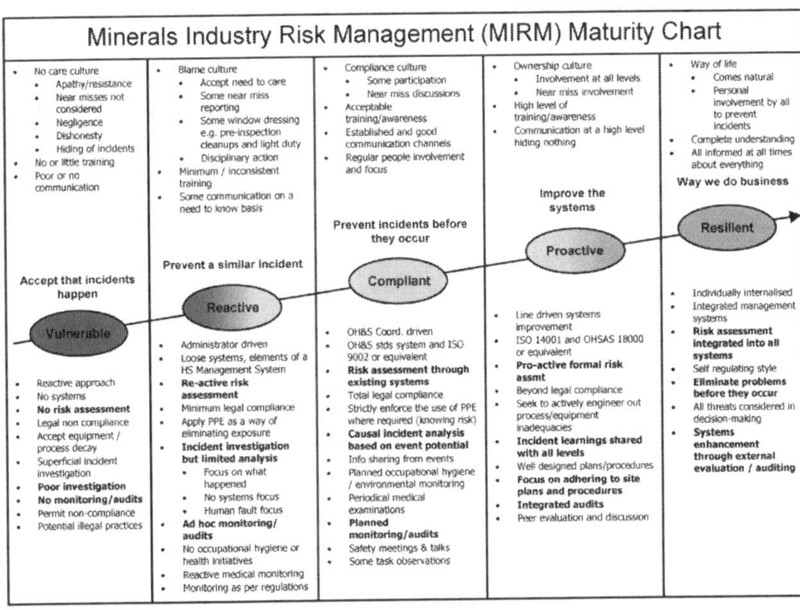

Figure 1. MIRM maturity chart [15].

In 2007, Anglo American plc embarked on a major new in-house initiative in order to revitalize and strengthen the companies approach to safety risk management and a major training programme was developed that was delivered to all their Managers worldwide [11,16]. As part of this training, the risk management process was explained and discussed as a "journey" based on the MIRM safety ladder that involves both people and systems [16]. Figure 2 shows their model depicted as an image of a spiral staircase that is used in all courses, with discussions and exercises to identify current status and plans to improve. Here assessments are made against 23 elements (6 people elements and 17 system elements).

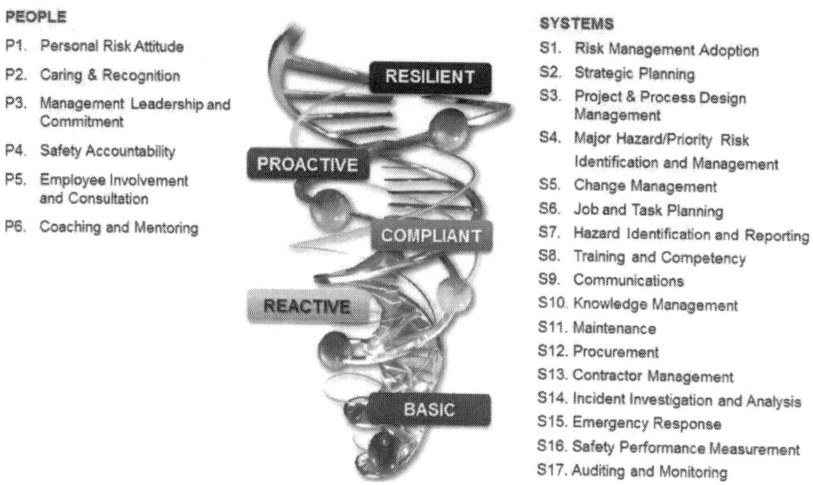

Figure 2. Anglo American Plc maturity model [16].

As part of this roll-out, all management teams have undertaken an assessment of where they are on the "safety journey" and plans have been developed for improvement that have been integrated into their operations.

It is important to note that these models are not just about assessing safety culture on its own. The Anglo Model, MIRM Ladder and Hudson Model all suggest a strong relationship between the culture of an organization and the development of a systems approach. It is critical to recognize that systems cannot progress up the ladder without culture progressing in parallel and *vice versa*. Gordon *et al.* [3] states that if there is a safety management system but no real commitment or culture towards safety, then the management system will not be effective, as decisions will not prioritise safety. Similarly, if there is a good safety culture, but no management system, then the way that safety is organized may be inconsistent, under-resourced and not seen as business driven [3]. It is possible that systems effectiveness issues that are currently experienced at mines sites are a result of such a culture/systems mismatch.

4. DEVELOPMENT OF A SAFETY MATURITY MODEL FOR UK COAL

4.1. The UK Coal Safety Way

By global standards UK Coal plc is a small mining company that took over many of the operating surface and deep mines when the UK coal industry was privatized in 1995. Although at the time of privatization it operated 20 deep mines and 27 surface mines in the UK, there has been a gradual reduction over the last 16 years to a point where it currently operates 3 deep mines and 6 surface mines. UK Coal remains the largest indigenous producer of coal in the UK, supplying approximately 5% of the total country's energy needs for electricity generation [17].

Although the mining industry in the UK is largely in decline, what has become increasingly obvious was the overall safety performance within the sector which appeared to be on a worsening trend. This period saw UK Coal's safety performance dip, from a period of almost seven years without a fatal accident, to a position where there were seven fatal accidents in deep mines between 2006 and 2009. This was accompanied by an increase in the level of significant incidents. In 2009 decisions were made to start and address the decline in safety performance, and in 2010 a new safety management system was introduced. This drew on best practices from other major mining companies, and other heavy industry. The revised management system was designed consisting of twelve standards (see Figure 3). Within each of the twelve elements a series of requirements had been formulated which needed to be adopted by all sites.

Whilst it was necessary to assess compliance against those standards, it was felt that any means of assurance should not just measure pure compliance as a "yes" or "no". It was necessary to incorporate a process that would give individual sites and departments targeted development areas to drive a continuous safety improvement down throughout the organization above and beyond pure compliance. It was felt that a "maturity model" approach would best suit this as it would highlight strong and weak areas and by considering it as a "journey", it would allow for continual improvement.

Figure 3. The UK Coal safety management system.

4.2. Development of the Methodology

It was decided to develop and use such a maturity model to measure progress in the integration of the safety management system standards at sites. Here, the "systems" and "culture" elements, that were separate in the Hudson, MIRM and Anglo American models, would be integrated within each of the 12 standards, which is unlike any of the safety maturity models described so far. The Maturity Model, which was named the UK Coal Journey Model, is shown in Figure 4.

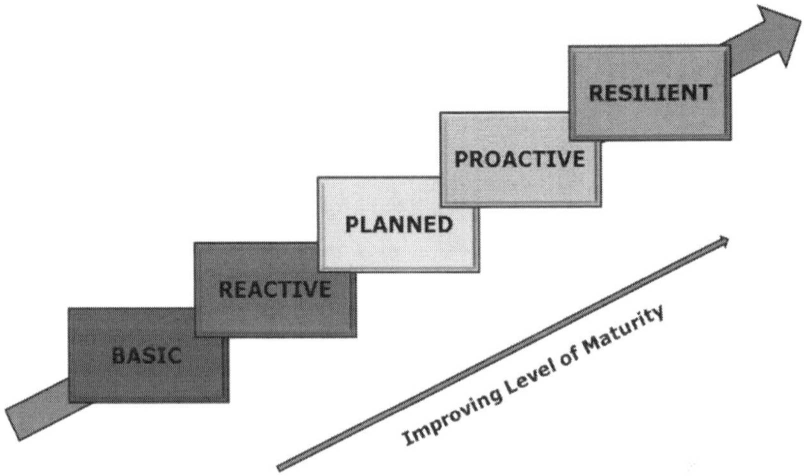

Figure 4. UK Coal journey model.

The model was developed so that the bottom level has few or no standards in place, and as a site moves up thorough the five levels they should start to see more of the requirements of the standards in place and there is improved compliance with, and effectiveness of these standards. There is also increased ownership of the standards by Manager, Supervisors and Operators. It should be noted that when examining the measurement system it was important that this could be applied at various levels within the organization. It was also important that there was not a sole reliance on the views of Management, but the process was also able to capture the Workforce views when forming a picture. This would come down to the way the model was applied at sites through team based "self-assessments".

When developing the model it was also felt that it was important to recognize that different sites will invariably be at different places on the UK Coal Journey Model, and it was equally possible that the same sites may be stronger in some of the areas covered by the standards than others. Therefore one of the significant differences here is that the steps are seen as overlapping on the model.

Table 2. UK Coal maturity model question set.

UK Coal Safety Way Element	Question Set
1. Leadership & Accountability	1. Safety Leadership & Commitment
	2. Safety Management Adoption
	3. Rewards of Good Safety Performance
	4. Work Planning
2. Policy & Commitment	1. Safety Accountability
	2. Safety *vs.* Production
	3. Safety Responsibilities
	4. Size & Status of Safety Department
3. Risk & Change Management	1. Major Hazard Risk Management
	2. Work Site JSA
4. Legal Requirements	1. Awareness of Legal Requirements
5. Objectives, Targets & Performance Measurement	1. Setting of Targets
	2. Monitoring & Accountability
6. Training, Competence & Awareness	1. Training & Competency
	2. Assessment of Training
7. Communication & Consultation	1. Communications
	2. Workforce Involvement & Consultation
8. Control of Documents	1. Document Control
9. Operational Controls	1. Jobs planning & Procedures
	2. Control Measures & Corrective Actions
	3. Maintenance
	4. Daily Inspections
10. Emergency Procedures	1. Standard of Emergency Planning
	2. Maintaining & Monitoring Response
11. Incident Investigation	1. Accident Investigation
	2, Quality of Investigations
	3. Follow up & Analysis
	4. Hazard & Unsafe Act Reporting
12. Monitoring, Auditing & Reviews	1. Safety Performance Measurements
	2. Monitoring & Auditing

The model was designed not to be an audit tool as such, but to take each standard, and look at how each standard is implemented and what occurs in practice. A self-assessment question set was developed to allow sites to determine their level of maturity (or how far they are "on the journey") by making observations for each standard at each of the 5 levels. From this, sites can look at what areas they need to improve, and then put together a specific action plan to help them move up a level.

As the model was starting to be developed it became very apparent following consultation with the senior management team with that the questions needed to be wide ranging in order to achieve the overall aim of each standard. As a result it was not possible to put together one description or one question per standard. To ensure that all of the implied requirements were met some standards in the model had to have up to four individual descriptions/question sets. The total number of questions/descriptions and their related subjects are shown in Table 2.

A separate "journey improvement" section was also documented, listing suggested actions and initiatives that could be undertaken for each element moving up each of the four stages ("Basic" to "Reactive", "Reactive" to "Planned", "Planned" to "Proactive", "Proactive" to "Resilient"). This was so that each Site could identify actions that they could implement, based on their findings, to move them up a level of maturity, or to move elements that were perhaps lower than the average, up to the average maturity level. As with any maturity model it is not possible for an organization to jump or skip a level, and also to realize that moving up a level can take some considerable time.

4.3. Trialing the Methodology

The descriptions and questions sets within the model were drafted and in order to test or "calibrate" the model an informal exercise was undertaken at one of the deep mines in January 2011. Following this some of the language and descriptions within the question sets were changed, but it was agreed that the overriding description for each element question set at each stage of maturity was accurate. This was then put together in a workbook so that it could be used at sites. A copy of the question set for Element 5 of the safety management system (Objectives, Targets and Performance Measurement) is shown in Table 3. This shows the two question sets relating to this standard covering (i) the setting of targets and (ii) monitoring and accountability. The overall requirements for this element are as follows.

1. Each site shall establish a measurable safety plan, quantifying corporate objectives and targets and site objectives and targets which shall be agreed by site personnel and corporate personnel. Objectives and targets shall be in line with the general target of enabling continual improvement of health and safety performance.

2. Measurement against the plan shall form part of the site and corporate safety accountability process.

3. Objectives and targets shall be communicated and understood by all appropriate personnel, including the Executive, Senior Management, line management, employees and contractors.

4. Adequate resources shall be assigned to ensure that the planned and agreed targets and objectives are met.

5. Senior management shall be issued with Safety Performance Indicators. Safety performance shall be part of those indicators to ensure that safety is a priority for management.

6. UK Coal shall ensure that objectives and targets are reviewed on a periodic basis to ensure they stay on programme and to agree changes when they do not.

7. Targets and objectives shall be both reactive—accidents, incidents and dangerous occurrences—and proactive—near hits, *etc.*and where appropriate, performance measures including benchmarking against established best practice.

8. The organisation and all sites will be responsible for the effective review of objectives, targets and performance indicators to ensure they remain relevant for the current safety risk.

The baseline studies at the three deep mines were undertaken in April 2011. There was some variation in the makeup of teams at sites with one site's management team initially undertaking the exercise, and at other sites teams consisting of managers, supervisors and operators were brought together. In some cases teams were formed that represented individual parts of the mine (such as the coal preparation plant and the production face).

Table 3. Example of question-set related to UK Coal safety way element 5.

Basic	Reactive	Planned	Proactive	Resilient
5.1. Setting of Targets				
Little evidence of safety related activities. Formal safety goals and objectives have not been identified, let alone documented.	Safety goals are based around improving safety performance, based on statistics from the previous year. This is set site-wide with a site plan established by the safety department that is passed around the senior management team, but not well communicated to the rest of the workforce.	Some safety targets related to improving standards or systems are undertaken, but actions for this mainly apply to the safety department. All targets are determined by the Safety department and sanctioned by Senior Management.	Senior Managers are involved in determining safety objectives in conjunction with the safety department	Every worker in the organization is accountable for specific risk control activities. Roles and activities are clearly defined for all levels in the site. Work teams independently establish their own work objectives.
5.2. Monitoring & Accountability				
There is no accountability as nothing has been set.	Accountability is weak and progress is rarely reviewed throughout the year but the safety department is held accountable for results.	Monitoring is carried out by the Safety department who also becomes accountable for actions and activities related to the safety management system. Accountability is split between safety and line management depending on the results.	Line managers and supervisors are held accountable for results. Safety initiatives/ activities are adequately resourced and action plans/ objectives are set and monitored.	The safety performance indicators are proactive, and a performance monitoring system is in place focusing on operational excellence.

The results from each of the exercises were collated and presented on a "spider diagram". Figure 5 shows two such diagrams from one of the mines (Mine A). Where there were differences in maturity noted within elements for different questions, a mean position was taken for the purpose of the diagram. In the diagram the acronyms S1 to S12 represent the 12 elements of the management system as listed in Table 2. Figure 5a shows the results of the original assessment undertaken by the Management Team at Mine A, and Figure 5b shows the combined results for a series of later studies undertaken by groups with Management, Supervisors and Operators within the same mine.

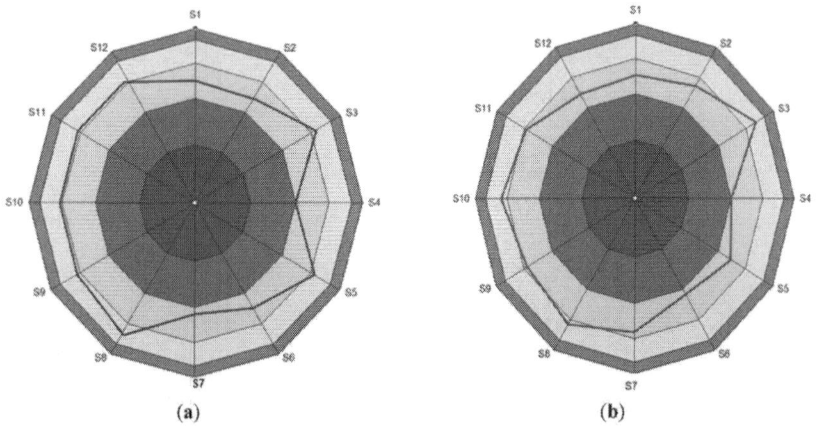

Figure 5. Spider diagram of results from Mine A.

It can be immediately seen that there is close mapping between the two in terms of weaker and stronger areas despite the studies being undertaken separately at different times and by different people. It can also be seen that there were no elements that were deemed to be resilient (the outer layer in the diagram). Elements 3, 5 and 8–12 were found to be at the "Lower Proactive" level with the remainder in the "Planned" level. This showed that the "doing" and "checking" part of the management system were strong, and that other elements (such as, 1. Leadership/Accountability and 2. Policy/Commitment) were weaker.

In terms of across the three deep mines as a whole, Table 4 shows the percentage of responses for each of the Management system elements across the five levels of maturity for the initial studies at the three sites (A, B and C).

It can be seen that some specific elements of the other two deep mines were seen to be "Resilient" by the groups undertaking the exercise. For example, 25% of the total responses across the sites put Element 3 (Risk and Change Management) and element 10 (Emergency Procedures) at this level. For Element 3, this was mainly down to the specific question set relating to "Work Site Job Safety Analysis" reflecting a well-established process at the mines to undertake task based risk assessments which would then be used to develop a method statement for the specific task. Although the hazards and control measures identified in these assessments were very generic, the mines had got into the process of regularly using these assessments or thinking about them before a task started. This was seen to be doing its job in reinforcing safety at the start of a specific routine task.

Table 4. Analysis of results from the deep mines.

Element	Basic	Reactive	Planned	Proactive	Resilient
1. Leadership & Accountability	-	8%	50%	38%	4%
2. Policy & Commitment	-	8%	50%	42%	-
3. Risk & Change Management	-	-	33%	42%	25%
4. Legal Requirements	-	33%	50%	17%	-
5. Objectives & Perf Measurement	-	-	58%	42%	-
6. Training, Competence & Awareness	-	-	75%	17%	8%
7. Communication & Consultation	-	-	42%	58%	-
8. Control of Documents	-	-	17%	83%	-
9. Operational Controls	-	-	33%	63%	4%
10. Emergency Procedures	-	-	17%	58%	25%
11. Incident Investigation	-	-	50%	37%	13%
12. Monitoring, Auditing & Reviews	-	8%	42%	50%	-

In terms of Element 10 (Emergency planning) this score emphasizes the detailed plans in place for dealing with the realization of major mining hazards. At one mine, this plan had been successfully used in late 2010 to successfully evacuate all the workers from underground following a significant ignition of methane gas in one of their gate roads.

By applying this type of process any areas of weakness across the company can be identified, and actions taken centrally to improve them. The individual sites can form plans to start improving their individual areas of weakness and, where there is overlap, work together. At Mine A improvement plans were produced by the Management Team after the initial study at to specifically improve elements 1, 2, 4 and 7 (all of which were in the "Planned" level) to bring them up to the mean of the 12 elements ("Low Proactive"). A summary of this plan is shown in Table 5.

The Journey Model has been used successfully to measure the current "baseline" as the new safety management standards are introduced. The intention is to run the exercise again across the surface and deep mine business with a view to informing the safety intervention strategy for the following year, both on site and as a Company, and to see whether there has been specific improvements from the initial study. In a drive to capture the broadest view and best inform the plan the exercise will continue be run with groups of Operational Managers, Supervisors and Site Operatives. The ultimate intention is to expand the use of this model, using it routinely for audit as well as developing it to measure individual business functions or even individuals within those functions.

Table 5. Action plan from Mine A.

Standard/Element	Required Changes	Specific Actions	As Measured by	Responsibility
1. Leadership & Accountability (1.3 Rewards of Good Safety Performance)	Promote/celebrate significant safety achievements.	Newsletter Screen in concourse.	Accident free stats Specific Work Standards Achievements	Safety Department Line Management HR Manager
2. Policy & Commitment (2.1 Safety Accountability)	Utilize front line supervisors to drive safety and "buy in" to safety.	Managers Weekly Meeting with Officials Engineers weekly meeting Safety Accountability meeting	Observation Auditing	Line Management
2. Policy & Commitment (2.4 Status of the Safety Function)	Clearer defined roles and responsibilities plus expectations within the mine management structure	HQ/Colliery Manager consultation Align to UK Coal Safety Policy	Compliance to Colliery Safety Action Plan	HQ/Colliery Manager
4. Legal Requirements (4.1 Awareness of Legal Requirements)	HQ Informing all line management in new/changes to legislation	Email line managers bulletin with changes to legislation Refresher updates for line management (away from pit)	Compliance to all legislation, company directives & codes & rules	HQ/Colliery Manager
7. Communication & Consultation (1. Communications)	Two way communication between management & workforce	Face & Development start up meetings All other meetings	Testing Knowledge Observations	Line Management

5. CONCLUSIONS

The Maturity model concept has already, through the work of Hudson and Fleming, been shown to be a useful tool for organizations to establish their levels of safety culture maturity.

The model here differs in that it is directly linked to a Company's own safety management standards and any cultural elements have been defined within this. It has been successfully used by teams of Managers, Supervisors and Workers to look at the maturity of the safety management system at their sites. Likening the maturity to a "journey" also helps make the model more practical and friendlier to use. Whilst ultimately sites would like to move forward together on the journey, it does recognize that different sites will be at different places, and it is often important at the early stage of any improvement that any weaker areas be targeted first.

ACKNOWLEDGEMENTS

The Authors would like to thank UK Coal and the work teams involved in the initial exercises at the mine sites. The views expressed in this paper are those of the authors.

REFERENCES AND NOTES

1. Fleming, M. *Safety Culture Maturity Model*; HSE Offshore Technology Report 2000/049: Sudbury, UK, 2000; pp. 3–7.
2. Hudson, P. Implementing a safety culture in a major multi-national. *Safety Sci.* **2007**, *45*, 697–722.
3. Gordon, R.; Kirwan, B.; Perrin, E. Measuring safety culture in a research & development centre: A comparison of two methods in the air traffic management domain. *Safety Sci.* **2007**, *45*, 669–695.
4. Keil Centre, *Managing Safety Culture in the UK Rail Industry: Report on the Review of Safety Culture Tools and Methods*; Rail Safety & Standards Board: London, UK, 2004.
5. Kyriakidis, M.; Hirsch, R.; Majumdar, A. Metro railway safety: An analysis of accident precursors. *Safety Sci.* **2012**, *50*, 1535–1548.

6. Lardner, R. Towards a Mature Safety Culture. In Presented at the Institute of Chemical Engineers Annual Conference, Manchester, UK; 2002.

7. Filho, A.P.G.; Andrade, J.C.S.; de Oliveira Marinho, M.M. A safety culture maturity model for petrochemical companies in Brazil. *Safety Sci.* **2010**, *48*, 615–624.

8. Fleming, M.; Lardner, R. Safety culture—The way forward. *Chem. Eng.* **1999**, *689*, 16–18.

9. Hudson, P. Applying the lessons of high risk industries to health care. *Qual. Saf. Health Care* **2003**, *12*, i7–i12.

10. Parker, D.; Lawrie, M.; Hudson, P. A framework for understanding the development of organisational safety culture. *Safety Sci.***2006**, *44*, 551–562.

11. *Reducing Error and Influencing Behavior*, 2nd ed; Health and Safety Executive: Sudbury, UK, 1999.

12. Changing Minds: A Practical Guide for Behavioural Change in the Oil & Gas Industry. Step Change in Safety Website. Available online: http://www.stepchangeinsafety.net/knowledgecentre/publications/publication.cfm/publicationid/16 (accessed on 14 November 2012).

13. Department of Resources, Energy & Tourism. *Leading Practice Sustainable Development Practice for the Mining Industry: Risk Assessment & Management*; Commonwealth of Australia: Canberra, Australia, 2008; pp. 15–16.

14. Joy, J. Minerals Industry Safety & Health Centre, University of Queensland. Personal Communication, 2011.

15. The University of Queensland, Minerals Industry Risk Management Maturity Chart. University of Queensland Minerals Industry Health & Safety Centre: Brisbane, Australia, 2008.

16. Detailed Journey Workbook, part of the A3 Safety Risk Management Process Training Material, Internal Document. Anglo American Plc: London, UK, 2010.

17. UK Coal Plc. *Annual Reports & Accounts 2011*, Available online: http://coalfieldresources.com/uploads./pdf/625_ukcoalplc2011bookmarked.pdf (accessed on 22 January 2013).

12

CHAPTER

Impact of Coal Mining on Self-Rated Health among Appalachian Residents

Shannon M. Woolley,[1] Ada O. Youk,[1] Todd M. Bear,[2] Lauren C. Balmert,[1] Evelyn O. Talbott,[3] and Jeanine M. Buchanich[1]

[1]*Department of Biostatistics, Graduate School of Public Health, University of Pittsburgh, 130 De Soto Street, Pittsburgh, PA 15261, USA*

[2]*Department of Behavioral and Community Health, Graduate School of Public Health, University of Pittsburgh, 130 De Soto Street, Pittsburgh, PA 15261, USA*

[3]*Department of Epidemiology, Graduate School of Public Health, University of Pittsburgh, 130 De Soto Street, Pittsburgh, PA 15261, USA*

ABSTRACT

Objective. To determine the impact of coal mining, measured as the number of coal mining-related facilities nearby one's residence or employment in an occupation directly related to coal mining, on self-rated health in Appalachia. Methods. Unadjusted and adjusted ordinal logistic regression models calculated odds ratio estimates and associated 95% confidence intervals for the probability of having an excellent self-rated health response versus another response. Covariates considered in the analyses included number of coal mining-related facilities nearby one's residence and employment in an occupation directly related to coal mining, as well as potential confounders age, sex, BMI, smoking status, income, and education. Results. The number of coal mining facilities near the respondent's residence was not a statistically significant predictor of self-rated health. Employment in a coal-related occupation was a statistically significant predictor of self-rated health univariably;

however, after adjusting for potential confounders, it was no longer a significant predictor.Conclusions. Self-rated health does not seem to be associated with residential proximity to coal mining facilities or employment in the coal industry. Future research should consider additional measures for the impact of coal mining.

1. INTRODUCTION

The Appalachian Region is a 205,000-square-mile area that follows the spine of the Appalachian Mountains and includes all of West Virginia and parts of 12 other states [1]. Residents of Appalachia and other rural regions in the United States have higher rates of poverty, lower education levels, and more limited access to health care [2]. Many Appalachian communities also bear the burden of environmental exposure to toxicants from coal mining [3], chemical industries [4], metal refineries [5], and environmental tobacco smoke [6]. Specifically, residents in Appalachia are at an increased risk for diseases with environmental components—such as heart disease [7, 8], cancer [9], diabetes [10], and obesity [11]—compared with other ethnic groups or those living in nonrural areas [12]. While numerous studies have confirmed that health disparities exist in Appalachia, there have been conflicting findings across studies as to whether these health disparities stem from the socioeconomic disadvantages found in the region and/or from environmental impacts [3, 13–17].

Coal mining is one of the major economic industries for eight Appalachian states (Alabama, Kentucky, Maryland, Ohio, Pennsylvania, Tennessee, Virginia, and West Virginia) [18]. The area's need and support for coal mining make it important to understand any potential health risks posed to those involved and those living nearby. Residents of Appalachian coal mining communities have expressed concerns regarding illnesses after reported exposure to contaminated air and water from coal mining activities [19]; however, it is unclear whether living near coal mining sites negatively affects health [15, 20]. Quantitative research on the relationship between residential proximity to coal mining sites and health consequences has been limited to studies in Great Britain and to a narrow range of respiratory illnesses. The studies found that residential proximity to coal mining sites was associated with elevated levels of particulate matter [21] and increased symptoms of respiratory morbidity [22].

The use of self-rated health (SRH) as a predictor of mortality is well-established [23]. Twenty-seven international and US-based community studies show impressively consistent findings for SRH as an independent predictor of mortality [24]. The purpose of this analysis is to investigate the impact of coal mining, measured as the number of coal mining-related facilities nearby one's residence or employment in an occupation directly related to coal mining, on SRH in Appalachia.

2. METHODS

2.1. Sample

The sample was drawn primarily from 10 counties of interest (five coal mining and five noncoal mining): coal: Boone, WV; Logan, WV; Mingo, WV; Monongalia, WV; and Raleigh, WV; noncoal: Berkeley, WV; Calhoun, WV; Tyler, WV; Yancey County, NC; and Cocke County, TN. A total sample of 9000 telephone numbers, drawn from a pool of all landline and cellular telephone phone numbers in the relevant area codes, was purchased from Genesys Marketing Systems Group (MSG) (http://www.m-s-g.com/web/index.aspx). Numbers were prescreened by Genesys MSG to remove nonworking or nonresidential numbers; 5054 numbers were available to be called, with a target sample size of 400–500, chosen as the target sample size because this was intended as a demonstration project.

2.2. Survey

The survey consisted of about 120 questions, collecting demographic, behavioral, and socioeconomic risk factors, attitudes about community and environmental satisfaction, access to health care, and physical and emotional health information. It was programmed in Ci3 (a questionnaire authoring application) and loaded into Win-Cati 5.0, a computer assisted telephone interviewing (CATI) program using the Windows platform. The Win-Cati system enabled supervisors to ensure that phone numbers were called on schedule and allowed for direct data entry, reducing the likelihood of data entry errors. Numbers were called at different times during the day, evening, and weekend call periods to increase the likelihood of contact, with an average of 2.2 call attempts to each number.

2.3. Statistical Analyses

SRH was measured using responses from the survey question, "Would you say that in general your health is...," rated on a 5-point Likert scale from 1 (excellent) to 5 (poor) (question 53). We independently compared SRH responses for groups expected to have different levels of exposure to coal mining. The first comparison was based on the self-reported number of coal mining facilities nearby that person's home or the number of "yes" responses to questions which asked participants whether or not they lived near the following facilities (questions 44–49): underground continuous coal mine, underground longwall coal mine, mountaintop removal coal mine, surface coal mine, impoundment pond, or coal preparation facility. For example, if someone lived near all of the possible coal mining facilities, they were considered to live near six coal facilities. The second comparison was based on the survey question, which asked participants, "Are you or have you been employed in an occupation directly related to coal mining?" (question 96). Graphical representations were created to compare and contrast SRH status for these questions. Because we were interested in SRH as related to coal mining, a predominantly male occupation, we examined the descriptive characteristics by gender and ran gender-stratified and not stratified models.

We used ordinal logistic regression to investigate whether the number of coal facilities nearby and/or having an occupation directly related to coal mining are predictors of SRH after adjusting for age, sex, and other potential confounders. Other potential confounders were included in adjusted multivariable models for SRH if the potential confounder had univariate statistical significance of . Due to the small number of females with work in coal mines, we restricted those analyses to males only. Score tests were used to assess the proportional odds assumption for all models. For the male only models, SRH was collapsed into three categories (excellent, very good + good, and fair + poor) in order to satisfy the assumption of proportional odds. Statistical significance for all multivariable models was assessed as . This study was approved by the University of Pittsburgh Institutional Review Board. Respondents were not provided an incentive for their participation.

3. RESULTS

Each telephone number in the sample was called at least once. Numbers called more than once were called back at different times during the day, evening, and weekend call periods to increase the likelihood of contact. An average of 2.2 call attempts were made to each number. A total of 415 interviews were completed. Using American Association for Public Opinion Research (AAPOR) standard definitions [25], the cooperation rate for the study was 34% (415/1235) and the response rate was 10.9% (415/3793), with an average interview length of 28.6 minutes. Figure 1 shows the distribution of responses for the outcome variable of interest, SRH. Approximately half of the respondents reported their health as excellent or very good.

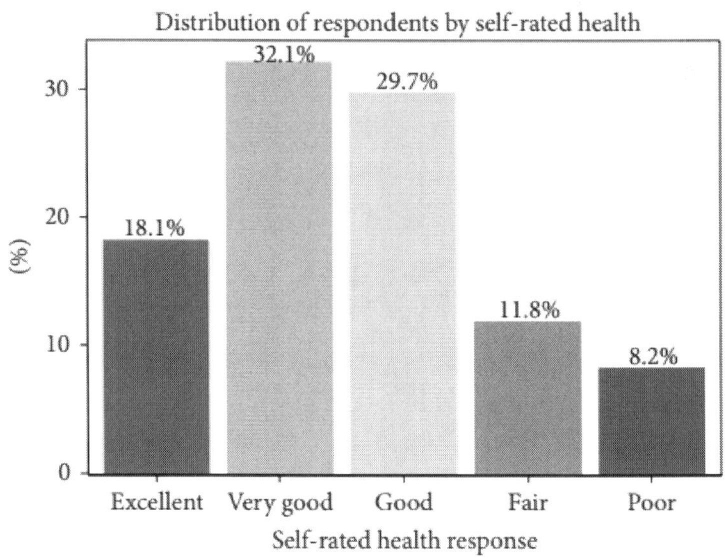

Figure 1

Table 1 shows, by sex, the demographic characteristics of the sample. Males were slightly more likely than females to be every day smokers, were less likely to earn less than $25,000, were more likely to have less than a HS degree education, and were more likely to live near a coal mining facility or be employed in a coal-related occupation.

Table 1: Descriptive statistics [mean ± SD or (%)] for covariates.

Covariate	Males (n = 195)	Females (n = 220)	Total (n = 415)
Age	54.0 ± 16.7	55.0 ± 17.5	54.5 ± 17.1
BMI	29.1 ± 5.7	27.8 ± 6.6	28.5 ± 6.2
Smoking status			
Not at all	151 (77.4)	169 (76.8)	320 (77.1)
Some days	9 (4.6)	17 (7.7)	26 (6.3)
Every day	35 (18.0)	34 (15.5)	69 (16.6)
Income level			
Under $25,000	42 (21.5)	70 (31.8)	112 (27.0)
$25,000 up to $49,999	42 (21.5)	46 (20.9)	88 (21.2)
$50,000 up to $74,999	43 (22.1)	39 (17.7)	82 (19.8)
$75,000 or more	68 (34.9)	65 (29.6)	133 (32.1)
Education level			
High school or less	90 (46.2)	80 (36.4)	170 (41.0)
Some college/associates	44 (22.6)	72 (32.7)	116 (28.0)
Bachelors or more	61 (31.3)	68 (30.9)	129 (31.1)
Number of coal facilities nearby			
0	82 (42.1)	144 (65.5)	226 (54.5)
1 or 2	33 (16.9)	32 (14.6)	65 (15.7)
3 or 4	36 (18.5)	26 (11.8)	62 (14.9)
5 or 6	44 (22.6)	18 (8.2)	62 (14.9)
Employed in coal-related occupation			
No	111 (56.9)	204 (92.7)	315 (75.9)
Yes	84 (43.1)	16 (7.3)	100 (24.1)

As shown in Table 2, age, BMI, smoking status, income, and education were highly statistically significantly related to SRH. As age and BMI increased, respondents were more likely to report poorer SRH. Lower incomes and less education were also negatively associated with SRH. Employment in a coal-related occupation was also statistically significantly negatively associated with SRH (). This relationship was more pronounced in males ().

Table 2: Ordinal logistic regression univariable models for association between SRH and covariates.

Covariate	OR	95% Wald confidence limits		*p* value
Age	0.98	0.97	0.99	<0.01**
BMI	0.92	0.89	0.95	<0.01**
Sex				0.77
Female	1.00			
Male	1.05	0.75	1.49	
Smoking status				<0.01**
Not at all	1.00			
Some days	0.81	0.39	1.65	
Every day	0.36	0.23	0.59	
Income level				<0.01**
Under $25,000	1.00			
$25,000 up to $49,999	1.65	1.00	2.73	
$50,000 up to $74,999	4.16	2.45	7.08	
$75,000 or more	3.77	2.36	6.04	
Education level				<0.01**
High school or less	1.00			
Some college/associates	1.35	0.88	2.07	
Bachelors or more	4.79	3.08	7.43	
Number of coal mines nearby				0.40
0	1.00			
1 or 2	0.94	0.57	1.54	
3 or 4	0.68	0.41	1.13	
5 or 6	0.75	0.45	1.23	
Employed in coal-related occupation				0.04*
No	1.00			
Yes	0.65	0.44	0.98	

* Statistically significant at $p < 0.05$.
** Statistically significant at $p < 0.01$.

Table 3: Ordinal logistic regression model for association between SRH, number of nearby coal facilities, and covariates.

Covariate	OR	95% Wald confidence limits		*p* value
Age	0.98	0.97	0.99	<0.01**
BMI	0.91	0.88	0.94	<0.01**
Sex				0.20
Female	1.00			
Male	1.29	0.87	1.92	
Smoking status				<0.01**
Not at all	1.00			
Some days	0.96	0.45	2.03	
Every day	0.37	0.21	0.64	
Income level				<0.01**
Under $25,000	1.00			
$25,000 up to $49,999	1.23	0.70	2.14	
$50,000 up to $74,999	2.73	1.52	4.90	
$75,000 or more	2.19	1.26	3.81	
Education level				<0.01**
High school or less	1.00			
Some college/associates	1.11	0.69	1.78	
Bachelors or more	2.52	1.51	4.21	
Number of coal facilities nearby				0.31
0	1.00			
1 or 2	0.91	0.53	1.55	
3 or 4	0.74	0.43	1.28	
5 or 6	0.60	0.34	1.06	

**Statistically significant at $p < 0.01$.

The number of coal mining facilities near the respondent's residence was not a statistically significant predictor of SRH. Adjustment by the covariates age, sex, BMI, smoking status, income, and level of education did not change the association between SRH and number of mining facilities nearby (Table 3). However, the model estimates suggest a potentially linear relationship, indicating that as the number of coal mining facilities nearby increases, the probability of poorer SRH increases. The effect of living nearby 5 or 6 coal facilities

Table 4: Ordinal logistic regression model for association between SRH, work in coal mining, and covariates (males only).

Covariate	OR	95% Wald confidence limits		*p* value
Age	0.99	0.97	1.01	0.23
BMI	0.90	0.85	0.95	<0.01**
Smoking status				0.01*
Not at all	1.00			
Some days	0.80	0.19	3.39	
Every day	0.27	0.11	0.65	
Income level				0.27
Under $25,000	1.00			
$25,000 up to $49,999	0.79	0.31	1.99	
$50,000 up to $74,999	1.84	0.73	4.61	
$75,000 or more	1.46	0.61	3.50	
Education level				0.03*
High school or less	1.00			
Some college/associates	2.23	1.00	4.95	
Bachelors or more	2.73	1.18	6.30	
Employed in coal-related occupation				0.98
No	1.00			
Yes	0.99	0.52	1.88	

*Statistically significant at $p < 0.05$.
**Statistically significant at $p < 0.01$.

versus living nearby 0 coal facilities approached statistical significance ().

In the stratified models, the number of nearby coal facilities was not statistically significant for males or females (data not shown). In the males only model, BMI and smoking were statistically significant predictors of SRH (); in the females only model, age, BMI, smoking, income, and education were all statistically significant predictors of SRH ().

As shown in Table 4, after adjusting for age, sex, BMI, smoking status, income, and level of education, employment in a coal-related occupation was no longer a statistically significant predictor of SRH in males (). Higher BMI, smoking every day, and having a high school or less education were statistically significant predictors of poorer SRH.

4. DISCUSSION

Compared to WV overall, the respondents of this survey were less likely to report their health as fair or poor (20% survey versus 25.2% WV) [26]; in fact, 50% of participants reported being in excellent or very good health. The percent of current smokers in this sample (22.9%) is lower than the WV average of 27.3% but still higher than the average for the US (19.0%) [26]. However, the median self-reported BMI () is considered overweight. These results are similar to those found by Griffith et al. (2011) that self-reported health status in Appalachia was incongruent with current health behaviors [27]. In addition, the covariate factors of the sample participants are representative of the entire study population [28]. More males than females reported living near any coal facility, although the overall percent (45.5%) provided adequate discrimination to use number of coal facilities as a covariate in the models. A much higher percent of males worked in a coal-related occupation; as such, these models were stratified to analyze males separately.

In the univariable ordinal logistic models, SRH is also associated with the demographic/socioeconomic factors of age, income, and level of education, as well as the personal risk factors of obesity and smoking every day. The number of coal facilities near the respondent's residence was not associated with SRH. Employment in coal-related occupations was found to be negatively associated with SRH.

As shown in Figure 2, there does not appear to be a trend for SRH category and number of coal mining facilities nearby; however, there are notable differences between the mine groups. The percentage of respondents who reported "excellent" or "very good" health decreases slightly as the number of coal facilities nearby increases.

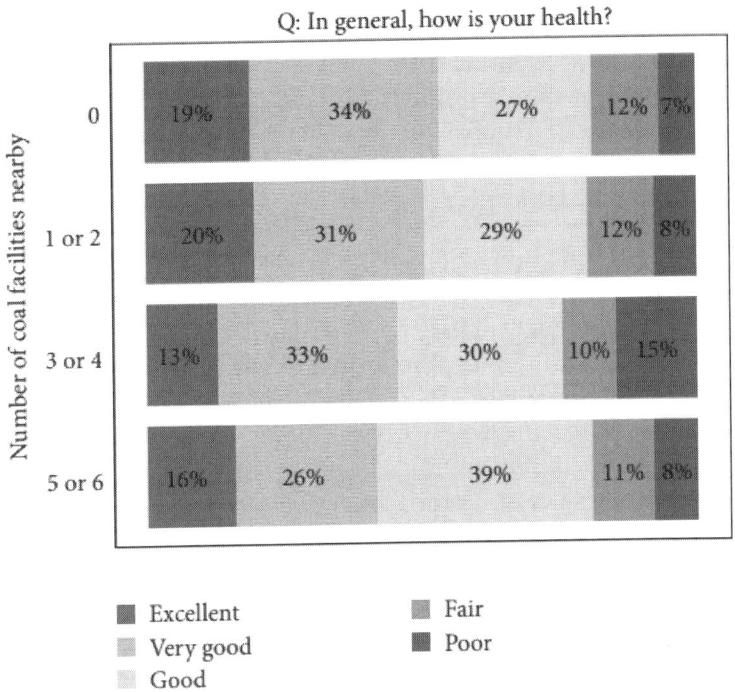

Figure 2.

Zullig and Hendryx (2011) found that residents of mountaintop mining communities generally had poor self-reported health [29]. In our survey, a separate analysis of questions regarding proximity to mountaintop removal and surface coal mines (questions 46 and 47, resp.) found no association between SRH and living near a surface coal mine (univariable ordinal logistic regression model, data not shown). Age, BMI, smoking, income, and education were all highly statistically significantly associated with poorer SRH in the multivariable model. These factors seemed to affect SRH in females more than males.

We found no evidence of an association between SRH and employment in an occupation related to coal mining. However, the percentage of participants reporting a "poor" health status was 9% higher for those who are or have been employed in an occupation related to coal mining versus those who are/have not. In addition, approximately 62% of men with mining employment have an education level corresponding to "high school or less," whereas the education levels of men without such employment are more evenly distributed. Thus, it is possible that, by including both of these variables in the model, the effect of coal-related employment on SRH is being masked by education level.

This study has some limitations. First, survey data can be prone to self-reporting bias, although SRH is a validated outcome [23]. The number of nearby coal mining facilities may not accurately reflect the effects of coal mining on the individual and, in some instances, respondents were unaware of the proximity of facilities. Future work should consider geospatial analysis using exact addresses and distances between coal mining facilities and residents' homes; however, these results are strengthened by the ability to control for individual-level covariates and exposures. After covariate adjustment, the results suggested a linear relationship between SRH and the number of coal mining facilities, but it was not found to be statistically significant. This could be due to the size of our sample. While we were able to detect statistically significant relationships between SRH and other covariates of interest, the number of interviews was relatively small. A larger sample size may be able to elucidate statistically significant relationships that this study was not, such as one between SRH and the number of coal facilities nearby. Additional studies should focus on the magnitude of this potential relationship using exact address information.

This study found no associations between SRH and residential proximity to coal mining facilities or employment in the coal industry. It did, however, provide support for the body of literature associating lower SRH with individual-level factors age, smoking status, income, and education.

DISCLAIMER

The views, opinions, and recommendations expressed herein are solely those of the authors and do not imply any endorsement by ARIES employees, other ARIES-affiliated researchers, or industrial members.

CONFLICT OF INTERESTS

The authors declare no conflict of interests.

ACKNOWLEDGMENTS

This study was sponsored by the Appalachian Research Initiative for Environmental Science (ARIES). ARIES is an industrial affiliates program at Virginia Tech, supported by members that include companies in the energy sector. The research under ARIES is conducted by independent researchers in accordance with the policies on scientific integrity of their institutions. Information about ARIES can be found athttp://www.energy.vt.edu/ARIES.

REFERENCES

1. Appalachian Regional Commission, Appalachian Regional Industry Report, 2014,http://www.arc.gov/images/appregion/AppalachianRegionIndustryReport2014.pdf.

2. G. Friedell, L. Linville, and S. Hullet, "Cancer control in Appalachia," Cancer, vol. 83, no. 8, pp. 1868–1871, 1998. View at Google Scholar

3. M. Hendryx and M. M. Ahern, "Mortality in appalachian coal mining regions: the value of statistical life lost," Public Health Reports, vol. 124, no. 4, pp. 541–550, 2009.

4. K. Steenland, S. Tinker, S. Frisbee, A. Ducatman, and V. Vaccarino, "Association of perfluorooctanoic acid and perfluorooctane sulfonate with serum lipids among adults living near a chemical plant,"American Journal of Epidemiology, vol. 170, no. 10, pp. 1268–1278, 2009.

5. E. N. Haynes, P. Heckel, P. Ryan et al., "Environmental manganese exposure in residents living near a ferromanganese refinery in Southeast Ohio: a pilot study," NeuroToxicology, vol. 31, no. 5, pp. 468–474, 2010.

6. M. E. Wewers, K. L. Ahijevych, M. S. Chen, S. Dresbach, K. E. Kihm, and P. A. Kuun, "Tobacco use characteristics among rural Ohio Appalachians," Journal of Community Health, vol. 25, no. 5, pp. 377–388, 2000.

7. E. Barnett, C. R. Williams, L. Moore, and F. Chen, "Social class and heart disease mortality among African Americans," Ethnicity and Disease, vol. 12, no. 4, pp. S3-76–S3-81, 2002.

8. J. A. Halverson, E. Barnett, and M. Casper, "Geographic disparities in heart disease and stroke mortality among black and white populations in the Appalachian region," Ethnicity & Disease, vol. 12, no. 4, pp. S3-82–S3-91, 2002.

9. P. A. Wingo, H. L. Howe, M. J. Thun et al., "A national framework for cancer surveillance in the United States," Cancer Causes and Control, vol. 16, no. 2, pp. 151–170, 2005.

10. J. Halverson, L. Ma, E. Harner, et al., An Analysis of Disparities in Health Status and Access to Health Care in the Appalachian Region, edited by L. Ma, J. Harner, Appalachian Regional Commission, Office of Social Environment and Health Research, Department of Community Medicine, Robert C. Byrd Health Sciences Center, West Virginia University, Morgantown, WVa, USA, 2004.

11. Centers for Disease Control Data and Statistics, Obesity, 2012, http://www.cdc.gov/DataStatistics/.

12. J. L. Hurley, H. S. Turner, and D. L. Floyd, "Development of a health service at a rural community college in appalachia," Journal of American College Health, vol. 48, no. 4, pp. 181–185, 2000.

13. J. Borak, C. Salipante-Zaidel, M. D. Slade, and C. A. Fields, "Mortality disparities in appalachia: reassessment of major risk factors," Journal of Occupational & Environmental Medicine, vol. 54, no. 2, pp. 146–156, 2012.

14. L. Esch and M. Hendryx, "Chronic cardiovascular disease mortality in mountaintop mining areas of central Appalachian states," Journal of Rural Health, vol. 27, no. 4, pp. 350–357, 2011.

15. M. Hendryx, K. O'Donnell, and K. Horn, "Lung cancer mortality is elevated in coal-mining areas of Appalachia," Lung Cancer, vol. 62, no. 1, pp. 1–7, 2008.

16. M. Hendryx, "Mortality from heart, respiratory, and kidney disease in coal mining areas of Appalachia,"International Archives of Occupational and Environmental Health, vol. 82, no. 2, pp. 243–249, 2009.

17. M. Hendryx, E. Fedorko, and A. Anesetti-Rothermel, "A geographical information system-based analysis of cancer mortality and population exposure to coal mining activities in west virginia, united states of america," Geospatial Health, vol. 4, no. 2, pp. 243–256, 2010.

18. F. Freme, Annual Coal Report, edited by: Department of Energy, Department of Energy, Energy Information Administration, Washington, DC, USA, 2005.

19. J. Goodell, Big Coal: The Dirty Secret behind America's Energy Future, Houghton Mifflin, Boston, Mass, USA, 2006.

20. M. Hendryx, M. M. Ahern, and T. R. Nurkiewicz, "Hospitalization patterns associated with appalachian coal mining," Journal of Toxicology and Environmental Health Part A: Current Issues, vol. 70, no. 24, pp. 2064–2070, 2007.

21. T. Pless-Mulloli, A. King, D. Howel, I. Stone, and J. Merefield, "PM10 levels in communities close to and away from opencast coal mining sites in Northeast England," Atmospheric Environment, vol. 34, no. 19, pp. 3091–3101, 2000.

22. T. Pless-Mulloli, D. Howel, A. King et al., "Living near opencast coal mining sites and children's respiratory health," Occupational and Environmental Medicine, vol. 57, no. 3, pp. 145–151, 2000.

23. M. Jylhä, "What is self-rated health and why does it predict mortality? Towards a unified conceptual model," Social Science and Medicine, vol. 69, no. 3, pp. 307–316, 2009.

24. E. L. Idler and Y. Benyamini, "Self-rated health and mortality: a review of twenty-seven community studies," Journal of Health and Social Behavior, vol. 38, no. 1, pp. 21–37, 1997.

25. American Association for Public Opinion Research (AAPOR), Standard Definitions: Final Dispositions of Case Codes and Outcome Rates for Surveys, AAPOR, 8th edition, 2015.

26. WV Health Statistics Center, West Virginia Behavioral Risk Factor Surveillance System Report, 2012, 2014.

27. B. N. Griffith, G. D. Lovett, D. N. Pyle, and W. C. Miller, "Self-rated health in rural Appalachia: health perceptions are incongruent with health status and health behaviors," BMC Public Health, vol. 11, article 229, 2011.

28. J. M. Buchanich, L. C. Balmert, A. O. Youk, S. M. Woolley, and E. O. Talbott, "General mortality patterns in Appalachian coal-mining and non-coal-mining counties," Journal of Occupational and Environmental Medicine, vol. 56, no. 11, pp. 1169–1178, 2014.

29. K. J. Zullig and M. Hendryx, "Health-related quality of life among central appalachian residents in mountaintop mining counties," The American Journal of Public Health, vol. 101, no. 5, pp. 848–853, 2011.

13

CHAPTER

Surface Coal Mining Methods in Australia

Rudrajit Mitra and Serkan Saydam

The School of Mining Engineering, The University of New South Wales, Sydney, Australia

1. INTRODUCTION

Minerals are one element of Australia's natural resource base. Other key natural resources include forestry and fisheries, and together with agriculture, they make a significant contribution to the Australian economy and Australia's Gross Domestic Product (GDP). Mining contributed 7.7% Gross Value Added in 2008-2009, up from 4.9% in 2005-2006 (University of Technology Sydney [UTS] and Monash University, 2010; Australian Bureau of Agricultural and Resource Economics [ABARE], 2010).

Australia's export earnings from energy and mineral commodities in 2010 were $A165 billion, 25 per cent higher than in 2009 (ABARE, 2010). Australia's black coal exports were worth almost $A43 billion in 2010, second only to iron ore ($A49 billion) which was Australia's biggest export earner for that year. The total production of raw black coal in Australia in financial year 2010 – 2011 was 405 million tonnes (Mt). Black coal occurs in all States of Australia and the Northern Territory, but New South Wales (NSW) with 46% and Queensland (QLD) with 38% have the largest share of Australia's total identified resources. Queensland (55%) and NSW (42%) produce the most black coal with locally significant operations in Western Australia, South Australia and Tasmania. Figure 1 shows the black coal locations in Australia.

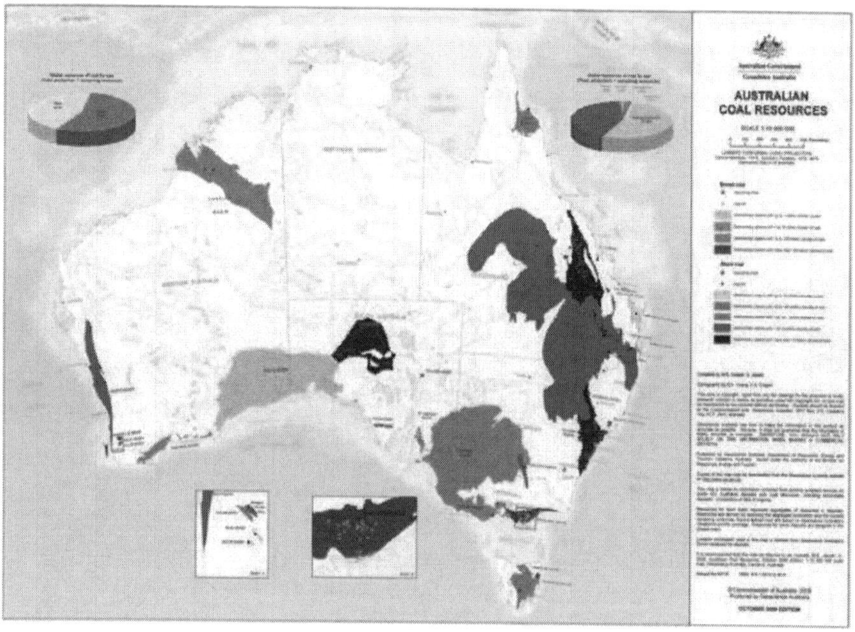

Figure 1. Australian coal resource map (Geoscience Australia, 2009).

Figure 2 indicates the Australian electricity generation by fuel in 2008/2009. The major uses of Australian coal are

- electricity generation – thermal or steaming coal
- steel industry – coking coal.

In Australia, the major coal mining methods are as follows:

- Underground mining operations:
    - Longwall mining, and
    - Bord and pillar mining (also known as room and pillar mining).
- Surface mining operations
- Open cut mining (strip mining)
    - Using draglines,

    - Using truck and shovels, and
    - Using a combination of above (integrated mining systems).
- Highwall mining
    - Continuous highwall mining, and
    - Auger mining.

This chapter will only focus on surface mining operations in the Australian coal industry.

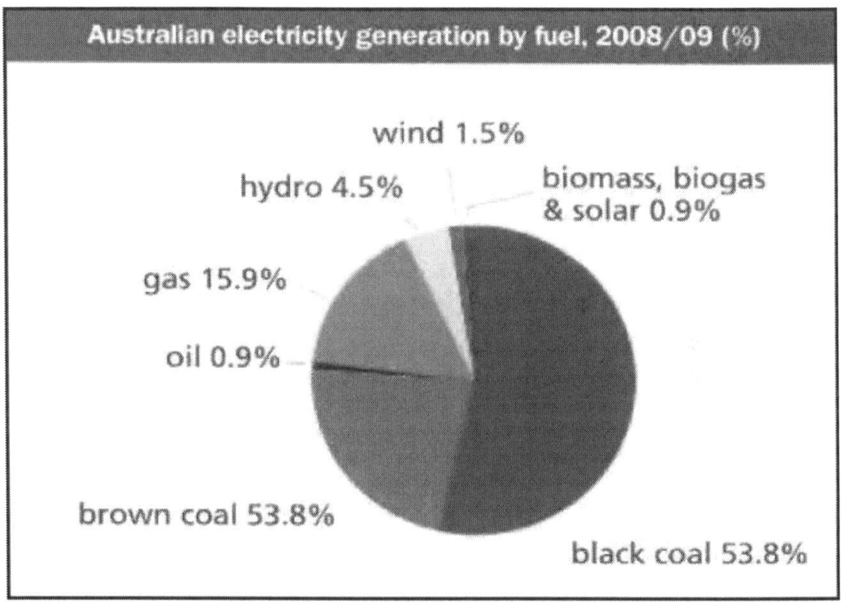

Figure 2. Electricity generation by fuel (ABARE, 2010).

2. OPEN CUT MINING METHODS – STRIP MINING

Large-scale open cut coal mining operations commenced in Australia in mid 1960s and since then there has been significant developments in this method of mining. The mines are now operating at significantly higher annual tonnages, growing deeper, more complex and operating at higher stripping ratios (Westcott et al., 2009).

The following major equipments are used in Australian coal mines:

- Draglines,
- Trucks,
- Shovels
- Bucket wheel excavators,
- Crushers and conveyors,
- Scrapers,
- Dozers,
- Slushers and dragline hoppers, and
- Surface continuous miners.

Initially open cut mines in coal mining was classified as strip mines where draglines were used, and open cut mines where truck and shovels were used. Currently a lot of mines use both equipment as integrated systems in Australia. Strip mining is ideally applied where the surface of the ground and the ore body itself are relatively horizontal and shallow and where a wide area is available to be mined in a series of strips (Westcott, 2004). Table 2 provides some examples of the Australian surface coal mines. Figure 3 illustrates a typical overall mining sequence of strip mine operation.

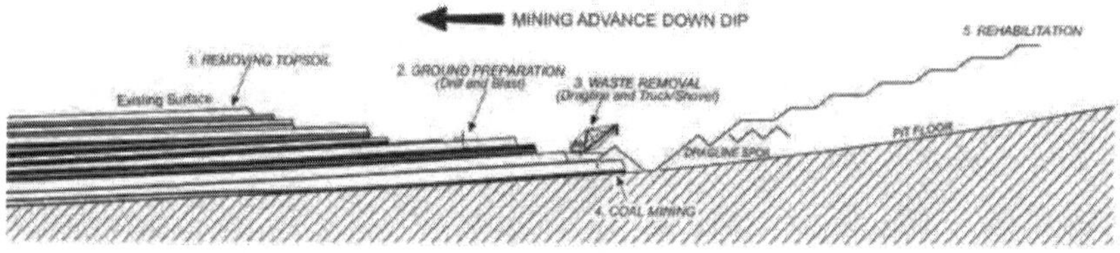

Figure 3. Overall mining sequence of a typical surface coal mine in Australia (after Westcott et al., 2009).

Table 1. Examples of Australian surface coal mines (adopted from Westcott et al., 2009).

Mining method description	Example mines
Multiple seam, steep dip, truck and loader, strike cut haulback mine with truck and loader coal mining.	Mt. Owen (NSW), Leigh Creek (South Australia), Duralie (NSW), Ravensworth East (NSW)
Several seam, moderate dip, dragline, down dip strip mine with throw blast, dozer assist and truck and loader prestrip.	Peak Downs (QLD), Norwich Park (QLD), Hail Creek (QLD), Curragh (QLD), Moura (QLD), Hunter Valley - Riverside (NSW)
Multiple seam, moderate dip, truck and loader, block cut, haulback mine.	Liddell (NSW), Wambo (NSW), Camberwell (NSW), Hunter Valley - Cheshunt (NSW)
Multiple seam, moderate dip, truck and loader, down dip strip, haulback mine.	Mt Arthur North (NSW), Muswellbrook Coal (NSW)
Thick, single seam, moderate dip, dragline - down dip - strip mine with throw blast, dozer assist and truck and loader prestrip.	Blair Athol (NSW), Newlands (QLD), Gregory (QLD), Goonyella/Riverside (QLD), Saraji (QLD)
Thick, single seam, moderate dip, bucketwheel excavator, block, cutback mine.	Yallourn (Victoria), Yoy Lang (Victoria), Hazlewood (Victoria)
Multiple seam, moderate dip, dragline down dip, strip mine with throw blast, dozer assist and truck and loader prestrip	Bengalla, Drayton, Bulga, Warkworth, Hunter Valley - Howick

2.1 Dragline method

Dragline is the predominant machine which is used to remove the overburden and expose the coal when the deposit's characteristics match the draglines' physical capabilities. Once the characteristics of the deposit alter from the physical limitations of the dragline, the overburden removal becomes costly as the machines' rehandle increases (Scott et al., 2010). In this situation, typically in multi-seam mines, truck and shovels are introduced to create an integrated mining sequence, which operates in tandem to generate a cost effective and efficient mining system (Fox, 2011).

Draglines are the lowest cost overburden removal equipment in common use. However, they are generally restricted to (Westcott et al., 2009):

1. Large deposits to ensure adequate strip length and sufficient reserves to justify the capital expenditure.
2. Gently dipping deposits, due to spoil instability on steep dips.
3. Shallow deposits, as draglines can only excavate a maximum of 50 to 80 m of overburden due to reach and dump height limitations. At greater overburden thicknesses draglines may

be supported by pre-strip with alternative equipment, but the compromises this entails usually adversely affects the cost effectiveness of the prestrip.

There were 68 draglines operating in QLD and NSW in 1992, with additional draglines either being built or ordered (Baafi and Mirabediny, 1998). While the most common machines average a 46m3 bucket size, there has been a trend in favour of larger machines. These draglines feature much larger bucket sizes (up to 115m3) and dimensions allowing the draglines to operate at greater depths and widths without needing pre-stripping and reducing rehandle (Aspinall et al., 1993; Gianazza, 2010). A typical dragline section is shown in Figure 4.

Draglines work in strips, which are typically 40 to 90 m wide and few kilometres long and the overburden is excavated by the dragline and dumped on the surface for the initial strip (box cut) and subsequently in adjacent mined out strips as illustrated in Figure 5.

Strips are generally aligned along strike with each subsequent strip down dip from the previous strip. The dragline starts at one end (or the middle) of the strip and advances along the strip to the other end. At the completion of each strip the dragline relocates to the start of the strip and commences the next strip, this is referred to as deadheading. Ramps for coal mining access are either taken through the overburden dump as a valley, dozed through the spoil pile parallel to the strip, created in the highwall either parallel or at right angles to the strip, or by a combination of these alternatives. Draglines are often supplemented by throw blasting and/or dozer assistance to increase overburden removal capacity (Westcott et al., 2009). Figure 6 illustrates a typical dragline while removing the material in one of the NSW mines.

The following are some of the advantages of using a dragline:

- Direct cast (excavate and transport),
- Low operating cost, and
- Handles hard digging.

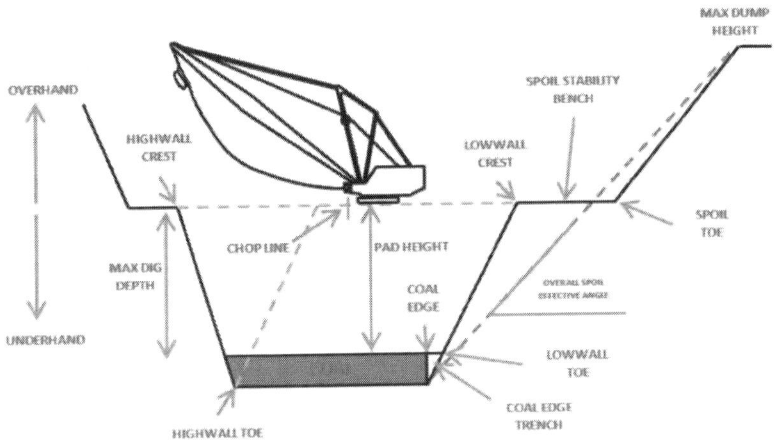

Figure 4. Dragline Working Section (Gianazza, 2010).

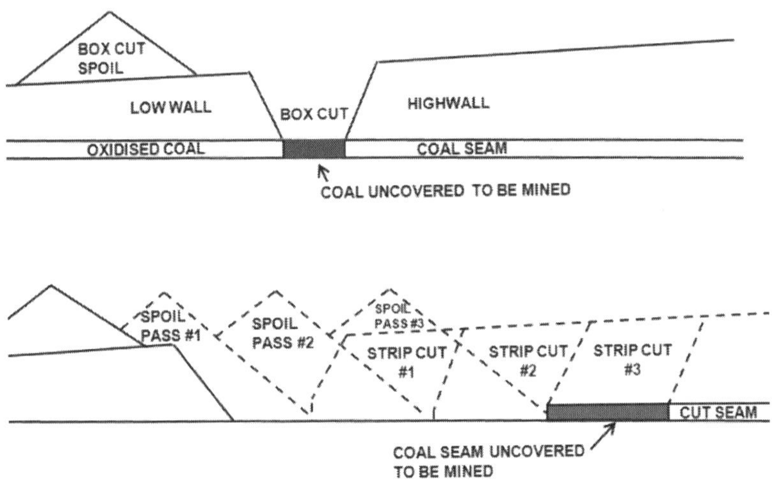

Figure 5. Box cut, strip cuts and spoil piles (Humphrey, 1984).

The following are some of the disadvantages of a dragline:

- Constraints on dig depth and dump height,
- Relatively inflexible,

- Requires detailed planning, and
- High capital cost.

The rigging on a conventional dragline weighs 20 tonnes and this has not changed in design in 100 years. This conventional rigging system limits the flexibility of the dragline operation and makes bucket control difficult. Along with high productivity costs when a dragline is out of action for rigging maintenance, there is also substantial safety hazards associated with replacing heavy pieces of equipment.

Figure 6. A typical dragline in operation in Hunter Valley Region in NSW.

A new technology was developed by the Cooperative Research Centre for Mining (CRCMining) Australia called the Universal Dig and Dump system (UDD). This replaces conventional rigging with a lighter, innovative configuration, improving operational flexibility. A medium sized UDD can move up to 13 tonnes more dirt in each pass and a specially designed computer system provides precise control over the bucket, enabling the dragline to dig and dump anywhere under the boom (Mining Technology Australia, 2011). This was developed in conjunction with BHP Billiton Mitsubishi Alliance (BMA). This has resulted in a significant improvement in productivity and is being installed on many draglines operating in Australia. Figure 7 shows one of these buckets in operation.

Figure 7. A UDD bucket in use (Mining Technology Australia, 2011).

2.2 Truck and shovel method

Truck and shovel mining method is the most flexible mining method and therefore better suited to geological complex deposits, varying overburden depths and thicknesses, and smaller deposits (Westcott, 2004). They offer cheaper capital investments than draglines, however they cost more to operate on a bcm (bank cubic metre) per hour removal rate. Another major benefit with the system is the ability to haul material long distance to ensure rehandle will not become an issue in the present time and future of the operation (Fox, 2011).

Truck and shovel method is inherently more flexible than dragline methods which makes them better suited in the following applications (Westcott et al., 2009):

1. Geologically complex deposits with resultant irregular pit shapes, which could not be efficiently mined by a dragline.

2. Steeply dipping deposits, where the equipment cannot operate on the seam roof and floor. Mining commences at one end of the deposit and advances along strike with strips laid out down dip. Overburden is initially dumped expit and then inpit when sufficient dump room is available. The pit is excavated as a series of horizontal benches (terraces) and coal and waste are exposed on every bench. Each bench extends from the floor of the lowest seam to the down dip economic pit limit.

3. Basin deposits that combine the problems of steep dips at the margins with short strike length and varying overburden depth along the strip.

4. Small deposits, which do not require the high productivities gained through use of a dragline.

'Haulback' mining is a particular application of a loader and truck operation. It refers to the fact that the waste is 'hauled' (transported) and dumped 'back' into the previously mined out area. Where the advance is down dip the configuration is not that different from a dragline strip mine and techniques such as throw blast and doze assist can also be incorporated. A good example of this design is

Figure 8. A view of a truck and shovel operation in Hunter Valley Region in NSW.

Mount Arthur North of BHP Billiton in the Hunter Valley, NSW. To keep haul distances as short as possible, many operations make use of rehandle bridges across the strip to transport waste from lower benches to the dump. As mining depth increases, a greater proportion of waste is long hauled around pit endwalls to the upper dump levels. 'Terrace mining' also dumps inpit but mines along the strike so the waste dump is keyed into a highwall. Multiple waste haulroads link the mining terraces to waste benches. This minimises haul cycle times for waste removal by reducing the vertical component of the haul. For the technique to work successfully both the mining terraces and the dump benches must be offset from adjacent benches to allow operational scheduling of mining activities. The method relies on the highwall being planned in its final position because once the mine is backfilled it would be a big economic hurdle to rehandle the waste to add another down dip increment. 'Cutback' open pit mining is where the waste is not dumped inpit but the waste is hauled to an external dump. Reasons for not dumping inpit could be that the floor is too steep for stable dumps or else the option of coming back later to either deepen the pit or extend the highwall down dip are part of the long-term plan. Panel shapes and sizes are dictated by production requirements, available equipment fleet and deposit geometry, rather than by parallel strips. Cutback mining may be used equally well in steeply or shallowly dipping deposits (Westcott et al., 2009).

The versatility of the system and ability to haul large distance makes this system extremely favourable in nearly all mining situations (Hays, 1990). Once equipment has been selected and the four main issues of cycle time are targeted and managed as best as possible to the conditions, will the system perform and achieve high production rates (Fox, 2011). A typical truck and shovel operation can be seen in Figure 8.

2.3 Integrated mining system - combination of truck & shovel and dragline

The integrated mining system where truck and shovel is combined with the use of a dragline for overburden removal, provides a system with many benefits. Generally, the truck and shovel system is used to remove the upper and thinner overburdens found within a deposit and the draglines remove the much deeper overburdens, which would be out of the working range of a shovel operation (Aiken and Gunnett, 1990), as shown in Figure 9.

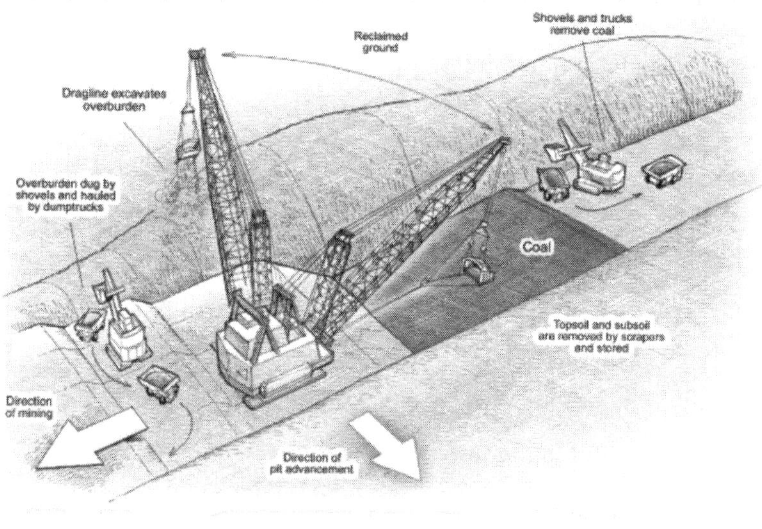

Figure 9. Integrated Mining System – combination of Truck & Shovel and Dragline methods (after Fox, 2011).

3. CASE STUDIES

3.1 Mount Thorley Warkworth mine, NSW

Mount (Mt.) Thorley Warkworth is an open cut coal strip mine located in the Hunter Valley, approximately 15km south of Singleton, NSW. The mine is owned by Coal & Allied and is operated by Rio Tinto Coal Australia, which is one of the largest mining companies in the world. Mt. Thorley Warkworth is a joint venture between Coal & Allied Industries Limited (80%) and Posco Australia Pty Ltd (20%). Coal & Allied has been around for 150 years and along with Mt. Thorley Warkworth. Figure 10 shows the location of Mt. Thorley Warkworth mine in the Hunter Valley Region.

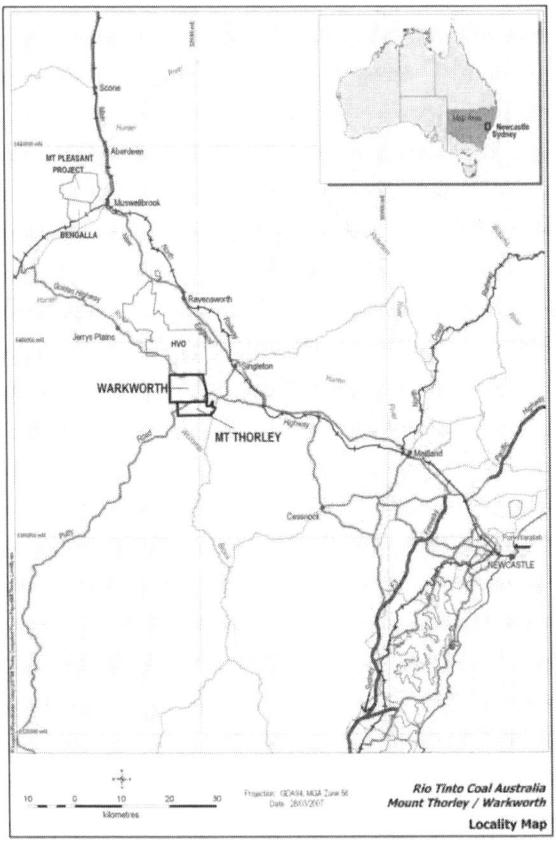

Figure 10. Mt. Thorley Warkworth locality map (after Fox, 2011).

Figure 11. Mt. Thorley Warkworth Mine aerial view (Fox, 2011).

Mt. Thorley Warkworth Mine currently produces around 12mtpa (million tonne per annum) of thermal export coal with the possibility of ramping up to 15mtpa in the future. The mine has operated an integrated mining system since the site's commencement in 1981. The site has two processing plants on site, due to the merger of the pits, and the coal type and quality depend on which plant the coal is delivered to. Once the coal is washed, it is then transported with the help of a conveyor to the coal loading unit for transport to Newcastle Port for exportation. Currently, the fleet system has 76 trucks ranging from 190t to 240t trucks, with the purchase of 330t trucks scheduled for 2012 (Corrigan, 2011).

The draglines operate in the lower portion of the seams where the interburden are the greatest and the pre-strip shovels and excavators are used on the upper portion of the deposit, where the interburden average is around 15-20m. The highwalls are 70 degrees, lowwalls are 45 degrees and the spoil and dump angles are 37 degrees. These angles allow the pit to operate safely as the geotechnical conditions are extremely favourable (Fox, 2011).

Mt. Thorley Warkworth Mine has a life of mine until 2050 and with the current coal processing capacities, the site has the ability to achieve a much higher production rate and increase the revenue generated. Figure 11 illustrates the aerial view of the mine and Figure 12 shows the truck and shovel system currently operating and removing the pre-strip material.

Figure 12. Mining sections in Mt. Thorley Warkworth Mine (Fox, 2011).

3.2 Curragh mine, QLD

Curragh Mine is located in the Central Queensland's Bowen Basin coal fields near Blackwater and operated by Wesfarmers Curragh Pty Ltd. The first successful extraction of coal occurred during the 1960's

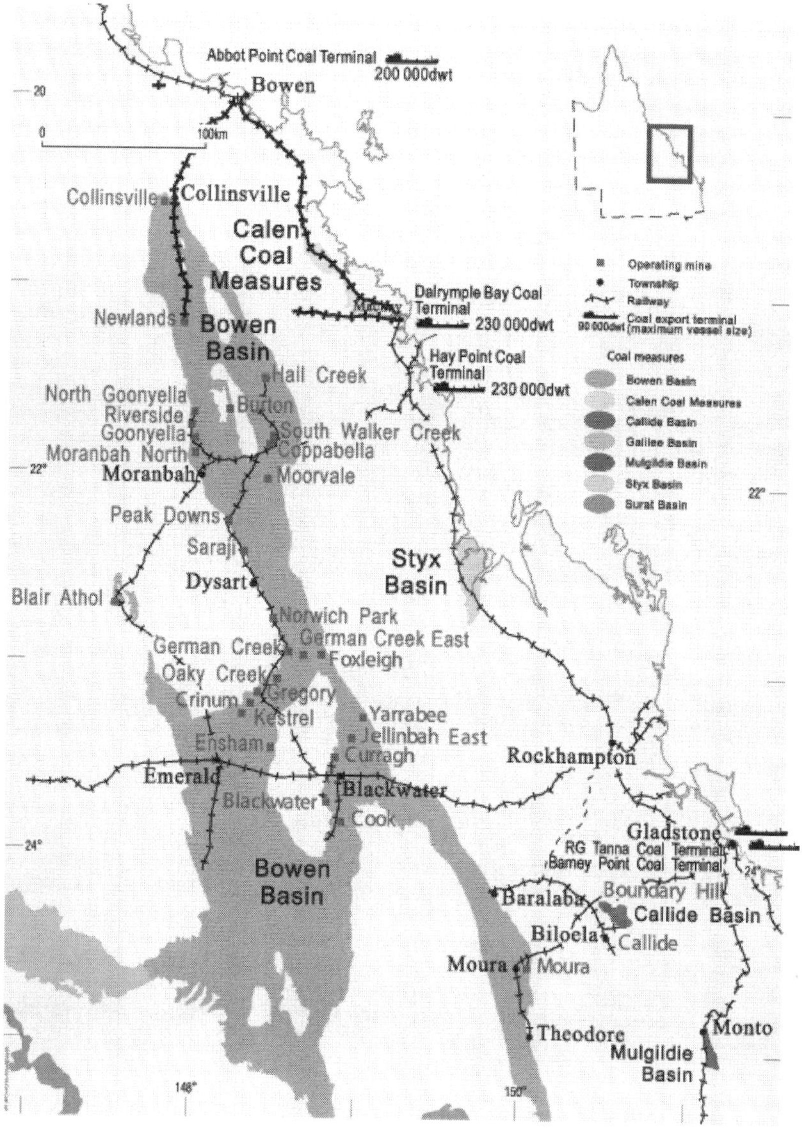

Figure 13. Location map of Curragh Mine (Gianazza, 2010).

in the area. In 1984, Curragh Mine began export of hard coking coal, mostly to Japanese and Korean smelters. Since then, the mine has undergone two major expansions: developing the Curragh East deposit on the eastern side of Blackwater Creek and the Curragh

North deposit, which more than doubled the recoverable reserves available in 2005 (Gianazza, 2010). Figure 13 shows the location of the mine.

Curragh Mine currently has a mine life up to 2025, with over 220 million tonnes of coal reserve within the mine plan and produces three different coal products: export metallurgical, domestic thermal and pulverised coal injection coals, each with varying specifications and parameters. The mine has a unique combination of complex geological settings and equipment compared with other mines in the Bowen Basin, QLD. Mining operations in Curragh Mine have five major seams with four of these seams mined in a single pit. The average seam thicknesses for each of the coal seams mined vary between 1 and 2.5m at the mine. Due to the geological complexity of the mine, the process differs significantly from pit to pit. Developed over a strike length of approximately 16 km, mining at Curragh Mine commenced at the line of oxidation (LOX) and has progressed down dip to the east. There is also a general trend in the topography sloping towards the east, limiting the increase in prime stripping ratio. Prior to removal of overburden, vegetation is removed, the land is assessed for indigenous significance in the area, and the topsoil stripped and stockpiled for later use in rehabilitating previously mined areas. With the exception of the first top 20m at Curragh North, all overburden and interburden are drilled and blasted for removal through either truck and shovel or dragline operations (Gianazza, 2010).

Curragh Mine operates four large capacity walking draglines as primary stripping machines, including one of the world's largest, with a 114m3 bucket. Depending on the pits and faulted areas, the draglines generally operate a multi-pass digging method, with strip widths usually in the range of 60-70 m range and highwalls designed at 63 degrees. The mine also operates a unique fifth dragline which provides significant operational flexibility, performing specific tasks more efficiently and effectively than a truck and shovel operation. This significantly facilitates overburden removal around highly faulted areas, due to its small size and high mobility. The coal removal process is undertaken by multiple fleets of hydraulic excavators, belly-dump coal haulers (204t) and rear dump trucks (180t). Coal is transported to the Coal Handling and Processing Plant (CHPP) and discharged at two 500t hoppers. The CHPP is dual line in design, with total crushing capacity of 2,400t of raw coal per hour. The plant allows flexibility by running coal to the preparation plant feed stockpiles through one line, and running bypass steam coal through another. Simultaneously processing two different coal types, the plant handles more than 1,400t of coal per hour through three modules. Product coal is stacked onto two stockpiles, each capable of holding 150,000t of coal. A bucket-wheel reclaimer loads the 7,000t coal trains at a rate of 3,000t per hour (Gianazza, 2010).

A total of 7 truck and shovel circuits combined with the operation of 5 draglines and the largest single span conveyor in the Southern Hemisphere, makes Curragh Mine both unique and successful at being one of QLD's leading producers of metallurgical and thermal coal (Brown, 2007). A monitoring system was placed in all 5 draglines in 2009, which provides production monitoring, high-precision GPS (Global Positioning System) and bucket depth control information back to operators.

3.3 Mount Arthur Coal mine, NSW

Mt. Arthur Coal Pty Ltd, fully-owned by BHP Billiton, operates the Mt. Arthur Coal open cut mine situated 5km south of Muswellbrook in the upper Hunter Valley of NSW (Figure 14). The mine produces thermal coal used for power generation. The coal is sold to both domestic and export markets. Production capacity of the mine is 20mtpa of raw energy coal and the operation shares the area with rural properties and other industries such as horse studs, vineyards, olive groves and residential suburbs.

Current mining activities are focused on 21 unique seams which were identified during the exploration stage. Mining is conducted using truck and shovel mining fleet. Mining equipment consists some of the largest and quietest trucks engineered in the world. The overburden mining fleet consists of rear dump trucks and rope shovels, while the coal mining fleet consists of trucks and a combination of excavators. Mining activities occur 24 hours per day, 7 days per week, on 12 hour shifts supported by approximately 1,000 full time equivalent personnel.

Mt. Arthur Coal currently operates under separate planning approvals for the Bayswater No. 3 Mine, Rail Loading Facility, Mt. Arthur North Mine, South Pit Extension, Exploration Adit and the Mt.

Arthur Underground (Figure 15). Open cut mining is conducted via a multi-bench, multi-strip shovel and excavator operations, which provide for the greatest operational flexibility and efficiency in the staged recovery of the coal resource at Mt. Arthur Coal. The extraction of coal via open cut methods is currently undertaken from the Mt. Arthur North and South Pit Extension area, as well as the Bayswater No 3 area, in which Saddlers Pit is mined using a contract fleet. To date only a limited amount of underground coal has been extracted from the Woodlands Hill seam via the Exploration Adit (Bailey, 2009).

Figure 14. The location map of Mt. Arthur Coal (Bill Jordan & Associates, 2009).

Large electric shovels are used to remove overburden and excavators are used to load coal into large capacity haul trucks for transportation to the CHPP and it is located adjacent to the mine.

Figure 15 shows an example of an excavator dumping onto trucks. Preliminary crushing takes place at the base of the hoppers before a conveyor transports the coal to the crushing station. Up to 2,000t of raw coal is fed through the crusher and 1,200t through the preparation plant per hour. The CHPP is where the coal is crushed, screened, cleaned and sorted according to market requirements. After the coal has been removed, rehabilitation is a priority at the mine. The overburden is replaced, shaped, covered with topsoil and replanted to resemble the original landscape (Umwelt, 2006).

Figure 15. Excavator removing the overburden and dumping onto trucks at Mt. Arthur Coal Mine (after Tredinnick, 2005).

Once crushed, the coal is transported to one of three locations:

- The domestic coal stockpile and then transported by conveyor to the local power station. Approximately 1.7 million tonnes of coal is used for domestic energy generation each year.
- The coal preparation plant where any impurities are removed, and then transported by conveyor to the export coal stockpile.

- Directly to the export coal stockpile and then to the export coal rail load-out facility for transportation to the Port of Newcastle. Approximately 10.7 million tonnes of coal was transported by rail to Newcastle in 2010.

The coal mining industry is concerned to minimise its impact on all aspects of the environment - visual landform, air and water quality, noise levels, native flora and fauna, soil conditions, and historic, indigenous and archeological sites (Australian Coal Association, 2011). One of the best examples of an environmental management of mining activities is Mt. Arthur Coal Mine. In 2010, approximately 12.3 million tonnes of coal was transported from Mt. Arthur Coal. From initial mine planning through to final site rehabilitation, the mine aims to minimise the impact of the operations on the natural environment and nearby residents.

A range of environmental issues and social impacts need to be managed at the mine site. The degree to which these are involved with the mine planning and operations department is determined by the location of the site and the inherent characteristics of the deposit and surrounding environment. Mt. Arthur Coal has a large number of statutory approvals relating to environmental management that cover the different activities on site. In total the operation must comply with more than 1,500 approval constraints, in addition to the requirements of government policies and legislation (Tredinnick, 2005).

The proximity of the Mt. Arthur Coal operations to the community of Muswellbrook exacerbates certain facets of the operation (Figure 16).

Effective management of these facets involves detailed mine planning. Some key facets of the operation, from a community perspective, that need to be addressed include (Tredinnick, 2005):

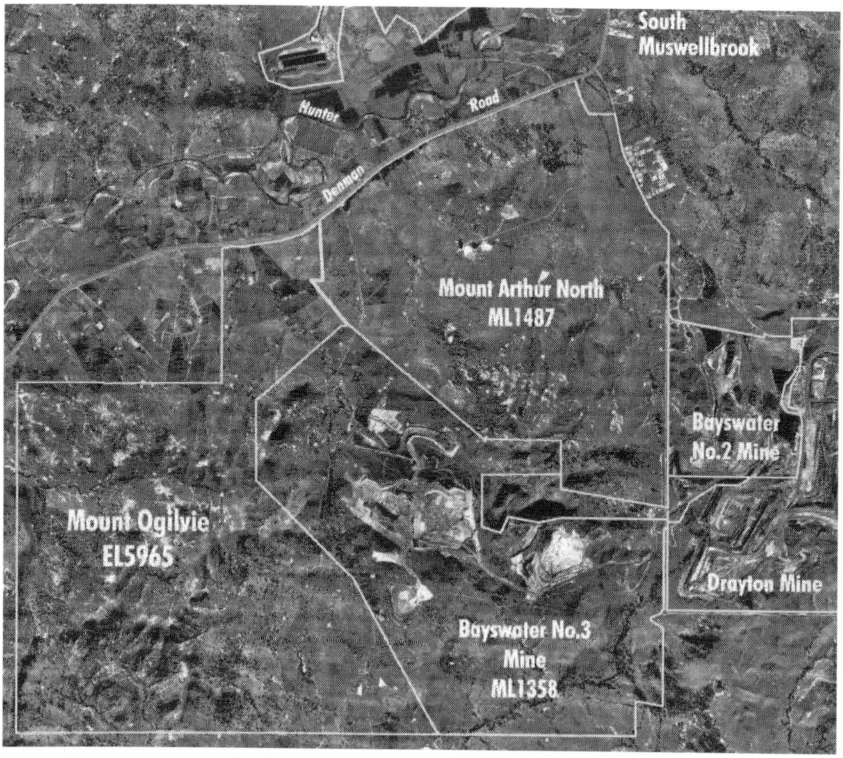

Figure 16. Mt. Arthur Coal Mine mining lease (after Tredinnick, 2005).

- Noise: Mt. Arthur Coal is required to meet some of Australia's most stringent noise limits for mining operations. These limits are set by government and are based on noise levels considered acceptable to the surrounding community. Haul trucks, dozers and excavation equipment are the most significant source of ongoing noise. The coal handling and preparation plant, and workshop areas are further sources of operational noise, although they have been located to minimise noise impact. From the first stages of mine design a noise model was developed to allow various mine designs and fleet configurations to be evaluated. Mt. Arthur Coal purchased a fleet of trucks that are the quietest trucks in the world. These trucks produce 16 times less noise output than a standard truck.

- Visual Amenity: The mine has an extensive visual impact management by doing progressive rehabilitation. To ensure safe working on night shift there is a requirement to illuminate working faces and overburden dumps. This use of lighting plants in turn has the potential to impact on nearby residents. There is an operational requirement to plan the location of all lighting plants to ensure that safe working conditions are maintained while at the same time minimising impacts on the local community.

- Blasting: The ground vibration and over pressure from blasting is strictly regulated. The 100% limits for Mt. Arthur Coal are a ground vibration of 10mm/s and an over pressure of 120 dBL. In addition, blasting is not allowed if the wind speed exceeds 10m/s. Actual ground vibration and over pressure results, as averaged across all monitoring sites, are approximately 0.3mm/s and 100dBL. In addition to ground vibration and over pressure NOx and dust generated by blasting are issues of concern to the local community.

- Spontaneous Combustion: Mt. Arthur Coal has committed considerable truck resources to long hauling clay rich overburden to the east pit to cap off areas of spontaneous combustion. This has seen over 95% of the spontaneous combustion eliminated.

4. CONCLUSIONS

Robertson et al. (2009) stated that the inexorable growth in world population and living standards places demands on coal resources for energy and metallurgical coke production. Within one more generation the global demand for food and energy will double. Coal is abundant and has many advantages: globally distributed, price is affordable and stable and readily converted to other valuable products. There is enough coal reserves for approximately next 120 years, which is almost 2 times more than gas and 3 times more than oil reserves. Coal provides 27% of global primary energy needs and generates 41% of the world's electricity. Australia has abundant high quality coal reserves and its geographical location close to the Asia-Pacific market has stimulated the development and expansion of the coal industry to become Australia's highest export earner. Geology, freight distances, political stability, established infrastructure and maturity enhance Australia's role in the supply of coal to world markets and Australia has vast resources of economically recoverable black coal reserves, capable of supplying demand for decades (Robertson et al., 2009). Australia is a world leader in surface mining technologies. In Australia, approximately 77% of black coal is produced from open-cut mines and surface mining equipment has increased dramatically in size over recent years. Recent developments are being made across various areas of the mining industry. With latest research and development achievements in automation, the surface coal mining productivity in Australia will continue to grow with lower cost and more environmental friendly.

REFERENCES

1. Australian Bureau of Agricultural and Resource Economics [ABARE] (2010). Energy in Australia, Australian Government, 13.12.2011,

2. Australian Coal Association (2011). Coal & The Environment - Coal Mining & Environment, 14.12.2011,

3. Aiken, G.E. and Gunnett, J.W. (1990). Overburden removal. In: Surface Mining, 2nd Edition, (Ed: B A Kennedy) pp. 584-619, Baltimore: Maryland, USA.

4. Aspinall, T., Isles, P. and Davidson, C. (1993). Strip Mining, In: Australasian Coal Mining Practice (Eds: A J Hargraves and C H Martin), pp. 210-220, The Australasian Institute of Mining and Metallurgy [AusIMM]: Melbourne, Australia.

5. Baafi, E.Y. and Mirabediny, H. (1998). Dragline digging methods in Australian Strip Mines – A Survey, In: Proceedings of Coal 1998: First Australasian Coal Operators' Conference (Ed: N Aziz), pp. 313-324, University of Wollongong and the AusIMM, Wollongong, Australia

6. Bailey, H. (2009). Mt Arthur Coal Consolidation Project - Environmental Assessment.

7. Bill Jordan & Associates. (2009). Appendix H - Blast Vibration Vulnerability Assessment.

8. Brown, N. (2007). Post blast evaluation using dragline diggability, Bachelor thesis (unpublished), University of Queensland, Brisbane.

9. Corrigan, M. (2011). Analysis of effect of carbon tax on operational costs of the Mount Thorley Warkworth truck fleet. BE Thesis. School of Mining Engineering, The University of New South Wales.

10. Fox, R. (2011). A Study on the effectiveness of mine design software for pit optimisation and scheduling at the Mt. Thorley Warkworth Coal Mine. BE Thesis. School of Mining Engineering, The University of New South Wales.

11. Geoscience Australia (2009). Australian Coal Resources.

12. Gianazza R. (2010). Strip optimization at Curragh mine. BE Thesis. School of Mining Engineering, The University of New South Wales

13. Hays, R.M. (1990). Trucks, in Surface Mining (2nd Edition), (Ed: BA Kennedy) pp. 672-691, Baltimore: Maryland, USA.

14. Humphrey, J .D. (1984). The Fundamentals of the Dragline, 29 p (Marion Division-Dresser Ind. Inc: Marion).

15. Mining Technology Australia. 2011.

16. Robertson, B., Ford, F., Osborne, D., Seib, W., Smith, G. andVan de Ven, P. (2009). The Australian Coal Industry Future Developments. In: Australasian Coal Mining Practice, Chapter 45, Monograph 12, Third Edition, Eds. Kininmonth, R.J and Baafi, E.Y. pp. 1057 - 1074, AusIMM, Melbourne, Victoria, Australia.

17. Scott, B., Ranjith, P.G., Choi K. and Khandelwa, M. (2010). A review on existing opencast mining methods within Australia, In: Journal of Mining Science, 46(3): 280-297.

18. Tredinnick, I. (2005). Environment & Community Opportunities and Challenges for Mine Planning and Operations, Presented at the School of Mining Engineering, University of New South Wales, May 2005

19. Umwelt Australia Pty Ltd (2006). Mt. Arthur Coal, Environmental Assessment Proposed Exploration Adit

20. University of Technology Sydney [UTS] - Institute for Sustainable Futures and Department of Civil Engineering Monash University. (2010). Peak minerals in Australia: a review of changing impacts and benefits 2010.

21. Wescott, P, 2004. The Inaugural UNSW/ Mitsubishi Lecture: Dragline or Truck and Shovel? - Some technical and business considerations. Presented to the School of Mining Engineering, University of New South Wales, May 2004.

22. Westcott, P., Pitkin, G. and Aspinall, T. (2009). Open-cut mining. In: Australasian Coal Mining Practice, Chapter 18, Monograph 12, Third Edition, Eds. Kininmonth, R.J and Baafi, E.Y. pp. 410 – 458, AusIMM, Melbourne, Victoria, Australia.

Index